1981

Life in
the Universe

AAAS Selected Symposia Series

Published by Westview Press
5500 Central Avenue, Boulder, Colorado

for the

American Association for the Advancement of Science
1776 Massachusetts Ave., N.W., Washington, D.C.

Life in
the Universe

The Ultimate
Limits to Growth

Edited by William A. Gale

AAAS Selected Symposium 31

AAAS Selected Symposia Series

Published in 1979 in the United States of America by
  Westview Press, Inc.
  5500 Central Avenue
  Boulder, Colorado 80301
  Frederick A. Praeger, Publisher

Library of Congress Cataloging in Publication Data
Main entry under title:
Life in the universe.
  (AAAS selected symposium ; 31)
  Includes bibliographies and index.
  1. Space colonies—Addresses, essays, lectures. 2. Astronautics and civilization—Addresses, essays, lectures. I. Gale, William A. II. American Association for the Advancement of Science. III. Series: American Association for the Advancement of Science. AAAS selected symposium ; 31.
TL795.7.L53   629.44'2   79-5132
ISBN 0-89158-378-5

Printed and bound in the United States of America

About the Book

This book posits an alternative to the pessimism of the phrase "limits to growth" by examining the prospects for extraterrestrial settlement. The authors, with backgrounds in space-program planning, planetary science, physics, political science, education, and futures studies, assume that considerable growth beyond the earth's surface is possible. They see a short-term prospect for the development of solar power satellites and electronic and biochemical production facilities in near-earth orbit; these would be followed by permanent human settlement in large, self-contained habitats. The resources in space would both support this growth and contribute to the standard of living of people remaining on earth, and the limits to growth would be encountered only when we reach volumes of space already developed by other beings.

About the Series

The *AAAS Selected Symposium Series* was begun in 1977 to provide a means for more permanently recording and more widely disseminating some of the valuable material which is discussed at the AAAS Annual National Meetings. The volumes in this *Series* are based on symposia held at the Meeting which address topics of current and continuing significance, both within and among the sciences, and in the areas in which science and technology impact on public policy. The *Series* format is designed to provide for rapid dissemination of the information, so the papers are reproduced directly from the camera ready copy provided, without copy editing. The papers are reviewed and edited by the symposia organizers, who become the editors of the various volumes. Most papers published in this *Series* are original contributions which have not been previously published, although in some cases additional papers from other sources have been added by an editor to provide a more comprehensive view of a particular topic. Symposia may be reports of new research or reviews of established work, particularly work of an interdisciplinary nature, since the AAAS Annual Meeting typically embraces the full range of the sciences and their societal implications.

WILLIAM D. CAREY
Executive Officer,
American Association for
the Advancement of Science

Dedicated to
Flavin Arseneau
whose teaching has inspired many

Contents

List of Figures and Tables

Acknowledgements

It is a pleasure to thank Lucinda Shastid for editorial assistance and for preparing the bulk of the manuscript.

Gregg Edwards originally suggested the idea of a symposium to me and provided support and encouragement at several steps since then.

The manuscript is set in Times Roman font by the Bell Laboratories computer typesetting system, TROFF.

About the Editor and Authors

William A. Gale, *a member of the Statistics and Data Analysis Research Department at Bell Telephone Laboratories, works on interdisciplinary problems in the fields of physics, economics, and statistics. Formerly a member of the technical staff of Bellcomm, a consultant to NASA, he studied Venus by radio astronomy. He is a member of the American Statistical Association and of the American Economic Association.*

Leonard W. David, *director of student programs at the Forum for the Advancement of Students in Science and Technology, specializes in the development of student programs on space-related topics. He is the author of numerous articles on space exploration and its applications, is a council member of the National Council on Aerospace Education, and has served as a consultant to the University Space Research Association and as a judge for the annual Robert H. Goddard Scholarship.*

Gregg Edwards *is program manager at the National Science Foundation for the program New Knowledge for National Productivity. He is Director of Professional Activities of the World Future Society, and has given many courses on futures methodologies.*

Michael A. G. Michaud, *U. S. Department of State, has published articles on interstellar negotiation, escaping the limits to growth, extraterrestrial colonization, and related topics. He has been State Department representative for the study of national space policy (1977-78), is a former editor of* Open Forum *(a classified opinion quarterly).*

Brian O'Leary, *research physicist at Princeton University, has worked both in planetary and space science and on long-term national energy policies. Formerly a scientist-astronaut with the NASA-Johnson Space Center, he has taught at various universities and has served as a consultant to the U. S. House Interior Committe. He is the author of many papers on planetary science and the potential of non-terrestrial resources for expanding* *the Earth's limits to growth, and he received the "Best Young Adult Book of 1970" award from the American Library Association for his book* The Making of an Ex-Astronaut *(Houghton-Mifflin, 1970).*

Jesco von Puttkamer, *program manager of Space Industrialization Studies in NASA's Advanced Programs Office of Space Transportation Studies, directs studies of advanced economic space activities and is responsible for NASA's long range program planning in space flight, particularly as it concerns permanent occupancy of space by humans. He is the editor of* Space for Mankind's Benefit *(GPO, 1972).*

Life in
the Universe

Introduction

William A. Gale

Are the terrestrial limits to growth the "Ultimate Limits to Growth"? The book, *Limits to Growth,* and the ensuing discussion has given this phrase a pessimistic connotation which may, however, be due to a lack of courage. Many sociologists believe that the long-range expectations of a culture have major and self-fulfilling expectations on its development. In an area with some, but limited knowledge, it is important critically to examine suggestions and assumptions. This symposium assessed the basic physical limitations to life support in the solar system, the galaxy, and the universe. The organization of the papers is roughly sequential in time of possible development of resources, starting with the longest period for which there is currently any government planning.

Jesco von Puttkamer discusses the long range plans of the National Aeronautics and Space Administration for the "Humanization of Space". He points out that the Space Shuttle Orbiter will introduce an age in which the use of space progresses from operational to routine. The key to this change is the lowering of transportation costs to space. Historically, lowering transportation costs to a new region has led to settlement of the region, and to an improved standard of living for the entire enlarged ecumene. While the use of space to date has been predominantly based on the synoptic view of the world that is provided from space, the commercial use of space has already begun. The provision of cheaper transportation costs to orbital space will increase the commercial opportunities many fold. In the coming decade, the largest growth in commercial use of space will be in telecommunications, followed by a growing space manufacturing industry.

Brian O'Leary reviews the engineering estimates for the exploitation of lunar and asteroidal resources which would follow the earth based exploitation of orbital space. The building of satellite power

1

stations from these materials may provide the cheap delivery to earth of a large quantity of electricity. Some asteroids are as cheap to exploit as the moon is, and can possibly provide large quantities of iron, nickel, and carbon in earth orbit at very low unit cost. It is possible that food could be grown in space-manufactured facilities for terrestrial consumption. It is projected that the entire world food supply could be produced in space if necessary. These resources could be used as soon as thirty years from now, O'Leary argues, but it is quite important to realize that they are continually available for development whenever we or our descendants choose to use them.

Leonard David adds a cautionary note that the rate of expansion in the use of space resources will be limited by conflict over resources in space, warfare extended into space, and pollution in space. The terms of current treaties governing the use of space have become outmoded by the advance of technology. A major current use of space is for military surveillance, which use, in the historical case of the airplane, was followed by the development of active and forceful interdiction of the surveillance space. A capability for warfare in space may be a deterrent to the commercial development of space. Furthermore, there is as yet no viable model of international resource development, and no court for the adjudication of conflicts. The position that resources not currently nationalized should be viewed as a "common heritage of mankind" expresses an incipient international conflict of goals. Planning for space activities has so far been deficient in calculations of the environmental impact. Although the space environment appears rich with desired materials and energy, there are also large amounts of less useful materials. Adequate disposal of these less useful materials is a possibly costly prerequisite to the continued safe use of space.

Gale and Edwards argue that that the ultimate limits to growth are set by the nearest other intelligent beings, and that that distance is likely to be a few billion light years. The previous papers show how the entire solar system can be exploited, with a continued phase of exponential growth for a few thousand more years. This paper argues that one way colonization trips to other other stellar systems at speeds between one hundredth and one tenth of the speed of light are now technologically feasible, and will become economically feasible to a solar civilization. Thus growth can continue, although its rate drops to a cubic expansion with time after the solar system is developed. Other galaxies are relatively closer for a galactic settlement than other stars are for a solar settlement. Therefore, we can expect to continue growing until regions are encountered in which the resources are already fully developed. It seems unlikely that the growth from this center is unique, so a probabilistic model of development and spreading of life forms from multiple centers is developed. The model suggests that the scale of time until it is likely that we will encounter another center of development is set by the time during which stars were forming and before the sun was formed, which is estimated to be five billion years. The observational consequences of this model are developed and stand in contrast to models which predict extraterrestrial intelligence within the galaxy.

Michael Michaud returns the discussion to the present with a discussion of how the prospects for life beyond the earth can be improved. He characterizes the problems as those of resource allocation, political decision, and societal direction, beyond the purely scientific research on feasibility. He suggests that the groundwork begins with an educational effort, and that an international set of institutions needs to be developed. We may conclude that laying the groundwork for the successful expansion of our descendants to the ultimate limits of growth is sufficient challenge for this generation.

1

Humanization Beyond Earth: The New Age of Space Industrialization

Jesco von Puttkamer

Abstract

Space flight will enter a totally new era with the availability of the Space Shuttle Orbiter. The new age will be distinguished by operational, nearly routine, use of the space environment. The key to this revolution is the reduction in transportation costs. The introduction of improved transportation has historically heralded improved living standards. Availability of transportation to space at dependable rates varying from a few thousand dollars to $31 million will permit the entrepreneurial development of the space environment.

The dominant aspect of space exploited so far is the synoptic view of the earth's surface and atmosphere. The global coverage available has improved services of weather, navigation, mapping, and telecommunications. Other features of orbital space that can be exploited profitably are free fall, solar energy, and high vacuum.

The commercial, profit making use of earth orbits has already begun. Commercial investments in satellites total over a billion dollars. Revenues from the communications satellite industry currently exceed $200 million per year. The provision of cheaper transportation to orbital space will increase the opportunities manyfold.

During the next decade, the largest growth in services provided from space will be in telecommunications. In the decade the beginning of space manufacturing is projected for products such as large perfect single crystals. This could be a billion dollar space market by the year 2000. Pure pharmaceutical products such as urokinase may be available most cheaply by space manufacture. The development of solar energy is a longer term prospect, potentially capable of providing 20% of United States power requirements for the year 2000.

-editor

Figure 1   Space Shuttle Approaching Landing

1. Introduction

With the dramatically successful completion of the Space Shuttle Orbiter approach and landing tests at NASA's Dryden Flight Research Center in November of 1977, space flight is now only one year away from entering into a totally new era. (Fig. 1)

While the main objectives of the National Aeronautics and Space Administration during the founding years of the U.S. space program, from 1961 to ca. 1975, were exploration, scientific discovery, capability development and pioneering "firsts" in space, the new age is distinguished by operational, nearly routine-like utilization of the cosmic environment for people on earth through space industrialization. With other words: If it frequently was the main thing in the past that space flight was undertaken at all, in future years it will increasingly be important *what* is being undertaken in space flight. The key to what amounts to a revolution in man's capability to bring the unique attributes of space into play to his advantage, is the reduction of transportation costs, i.e., a space transportation system more economic and cost-effective than the "throw-away" rockets of the past.

The introduction of improved transportation, be it wheeled ox-carts, more seaworthy ocean-going vessels, better motorized vehicles or faster airplanes, has always in history heralded a dramatic rise in standards of living in all affected areas of the world. Economical reusable space transport will undoubtedly have a much stronger impact on our world, by supporting and fostering consecutive and expansive exploration, scientific experimentation, commercial processing and services, as well as a large variety of beneficial satellite and manned space station applications.

But to be of maximum utility to mankind the principal goal of the space program of the 80's must be to improve the "quality of life" on earth. Planning for future activities in space, thus, must primarily focus on those basic themes which promote social purpose and progress, environmental protection, international understanding, politics, and which have the public's interest. The Space Transportation System (STS) is the backbone of such a practical and responsible long-range program of progressive utilization of space that leads from the initial attainment of easy access to and from space, ushered in by the Space Shuttle, to the industrialization and concomitant humanization of space, made possible by the eventual establishment of permanent space occupancy by man.(1)

Figure 2 Space Shuttle in Flight

2. The Space Transportation System

The basis for NASA's planning for Space Shuttle operations in the decade ahead is the so-called STS Traffic Model which currently lists a total of 487 Shuttle missions between 1980 and 1992, based on a five-Orbiter fleet and two launch sites (Eastern Test Range at Kennedy Space Center, Florida, and Western Test Range at Vandenberg AFB, California). During that period, up to almost 60 flights a year are scheduled to satisfy the various governmental, academic, commercial and foreign users wishing to avail themselves of the Shuttle's projected versatility, reliability, reusability and economy. (Fig. 2)

Each flight might be used either for only one specific payload, such as a large free-flyer spacecraft, or for a number of smaller payloads, satellites, etc., limited only by volume, weight, and some other physical characteristics. Such multi-cargo flights will provide each customer with the advantage of cost sharing. Transportation charges are anywhere from less than $10,000 to $31 million, depending on cargo length, weight, destination, user class, whether it is on a reservation or stand-by basis, and what optional services are desired beyond the standard services covered by the basic charge. (3)

A fare of $31 million will book the full capacity of the Shuttle on a reservation basis by non-U.S. Government customers, with 60 ft (18 m) by 15 ft (4.50 m) of cargo bay and 65,000 lbs (30 t) of payload, including flight planning, three-man flight crew, one day of on-orbit operations at standard mission orbit, and deployment of a free-flyer. At the low end of the charge structure, NASA offers a special "cut-rate" fare for small payloads not exceeding 200 lbs (90 kg) mass and 5 cubic feet (.14 m^3) volume. This class of small self-contained payloads, dubbed "get-away specials," permits an individual or organization (e.g., sponsor of an educational institution) to fly a research payload in the Shuttle on a space-available basis for $3,000 to $10,000, depending on package size. There are reduced-price incentives for payloads having exceptional merit, and added charges for short-term callups, postponements and cancellations. Lower rates are offered for stand-by payloads, for floating launch date options, and for reserving space on future missions.

Major flight elements of the STS, in addition to the Shuttle, are the Spacelab, a number of different upper propulsion stages, such as the Inertial Upper Stage (IUS), and various free-flyers, e.g., the Long-Duration Exposure Facility (LDEF), and the Power Module.

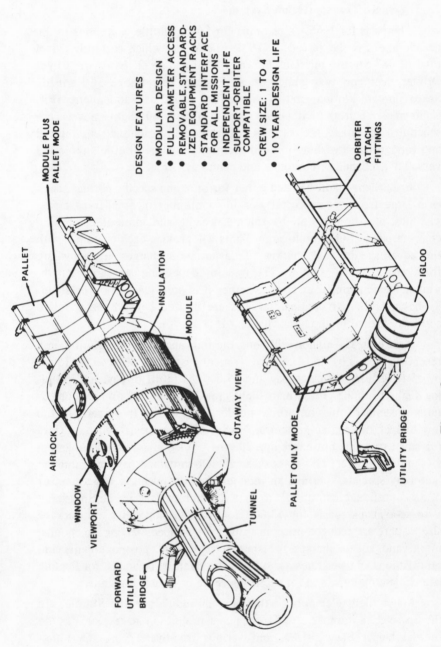

PALLET

MODULE PLUS
PALLET MODE

DESIGN FEATURES

● MODULAR DESIGN
● FULL DIAMETER ACCESS
● REMOVABLE,STANDARD-
  IZED EQUIPMENT RACKS
● STANDARD INTERFACE
  FOR ALL MISSIONS
● INDEPENDENT LIFE
  SUPPORT-ORBITER
  COMPATIBLE

CREW SIZE: 1 TO 4
● 10 YEAR DESIGN LIFE

ORBITER
ATTACH
FITTINGS

INSULATION

MODULE

CUTAWAY VIEW

AIRLOCK

WINDOW

VIEWPORT

TUNNEL

FORWARD
UTILITY
BRIDGE

PALLET ONLY MODE

UTILITY BRIDGE

IGLOO

Figure 3 Spacelab Design Concept

Figure 4 Power Module

Spacelab, currently under development by the European Space Agency (ESA), is a fundamental part of the system. It complements the Shuttle by allowing scientists and technicians to perform orbital experiments in a shirt-sleeve laboratory environment. A similar capability had existed in 1973 on board NASA's experimental space station "Skylab," but Spacelab has the advantage of being reusable and reflyable -- up to 50 times (or 10 years lifetime) -- in many different configurations and outfittings for numerous scientific-technical applications. It will be less expensive to operate and will permit a wide variety of experiments which can evolve from mission to mission. On its first flight in 1981, one of its male or female crew members, called "Payload Specialists," will be a European. (Fig. 3)

The Power Module, under design to enter operations in 1984, will augment the STS by supplying additional on-orbit sun-generated energy (about 25 kW, average), thermal heat rejection, and control-moment-gyroscope-driven attitude control to the Orbiter, extending its maximum orbital staytime of 30 days (at 7 kW fuel-cell power) to sixty or more days (at 11 kW), after which food, water, and other crew living factors limit duration.(4) The Power Module is also being studied as an orbiting energy/utilities facility to support high-power free-flying payloads detached from the Orbiter, and space-based construction operations. (Fig. 4)

In addition to performing the tasks planned under the STS Traffic Model, the Shuttle can be expected to take on new and as yet unrecognized space missions that will make the unique attributes of space increasingly accessible to our grasp and amenable to industrial use.

3. Definition of Space Industrialization

Space industrialization is the overarching long-range concept of progressively using the characteristics of space in satisfying commercial and utilitarian purposes, in contrast to -- but certainly not to the exclusion of -- scientific research and exploration. It entails all activities necessary to bringing this utilization to a level of operation where new values, products, goods and services become available which companies can sell at a profit, or those public services, for which citizens, through their taxes, are willing to pay. It is entirely possible that it will be the very commercialization of space flight that will provide a stable guarantee for longer-term, if not permanent, activities of humans in space. (Fig. 5)

WHAT IS SPACE INDUSTRIALIZATION?

SPACE INDUSTRIALIZATION IS NOT ...

- A "PROGRAM".
- A SPACE STATION.
- SPACE COLONIZATION.
- LARGE SPACE STRUCTURES.
- NEW OR 15 YEARS AWAY.
- A SPECIFIC ACTIVITY.

SPACE INDUSTRIALIZATION IS...

- INDUSTRY AND GOVERNMENT WORKING TOGETHER FOR PROFIT AND PRAGMATIC BENEFIT UTILIZING SPACE

Figure 5   What is Space Industrialization?

Seen from the prospect point of the long-range planner, the industrialization of near-earth space promises to become the focal point of a third industrial revolution which, in the far term, will supply not only products and services from space but also energy from the sun and resources from the moon, and which eventually will "humanize" the cosmic environment for it to become the natural habitat of people.(2)

4. Useful Attributes of Orbital Space

The dominant aspect of space exploited so far is the wide view of the earth's surface and atmosphere afforded by it. The earth resources satellite LANDSAT, for example, using multi-spectral scanning from 570 miles altitude, surveys 13,225 square miles of surface *every 25 seconds.* It is this global and synoptic coverage that is letting applications satellites prove their worth in the fields of telecommunications, weather, navigation, mapping, and studying earth's resources.

The fundamental characteristic of orbital flight most important to industrial use of space is the phenomenon of free fall where gravitational effects are largely absent and weightlessness reigns. This unique condition not only makes possible physical processes that are impossible to carry out on earth or, at the best, hampered and disturbed by gravitational effects, but also allows the construction of very large structures, reflectors, optics, telescopes, and antennas for information transfer and data acquisition, energy collection and beam-power transmission space-to-space and space-to-ground -- filigreed superstructures that would collapse on earth under their own weight.

For mankind at large, perhaps the most consequential feature of space is the essentially uninterrupted supply of solar energy for light, heat, and power. More about this later.

Other vital characteristics are the absence of an atmosphere and the attendant extremely high vacuum. The former enables us to make observations of earth and the universe with unparalleled coverage and distortion-free clarity. The latter means we can carry out processes, which on earth would require expensive and volume-limited vacuum equipment, pumps, etc. The acoustic isolation afforded by the near-perfect vacuum, combined with the seismic isolation of a freely coasting body, can be of crucial importance to certain research investigations and industrial processes.

Another unique and useful attribute is the unlimited reservoir that space provides for removal of waste heat and for storage and disposal of industrial waste products. In addition, the isolation from earth could prevent environmental damage by hazardous industrial processes and permit activities which are potentially harmful to earth's biosphere, such as energy- and pollution-intensive processes, certain genetic-engineering research programs (e.g., recombinant DNA experimentation), or other activities requiring special quarantine.

5. Goals of Space Industrialization

Major developments in the industrialization of space will focus on four main areas: information services, products "made in space," energy from space, and -- in the longer run -- people industries and space settlements.(2)

Space industrialization, i.e., the commercial, profit-making use of earth orbits, has already begun. Outstanding examples are the communications satellites, which are revolutionizing world-wide telecommunications.(5) Commercial investments in these satellites total well over a billion dollars. The Communications Satellite Corporation (COMSAT), a joint government-industry corporation based on private enterprise, returned its first dividend to its stockholders in the fourth quarter of 1970, six years after the initial stock offering, and has consistently paid a dividend ever since. Its net income after taxes during the third quarter of 1977 was $9.5 million. In the fourth quarter of 1977, the dividend was increased from $.25 a share to $.35. The revenues from the communications satellite industry currently exceed $200 million per year and are expanding rapidly.(6) This new industry is expected to be a multi-billion dollar one by the mid-1980's. Other examples are weather satellites, and satellites for navigation, traffic control, mapping and earth resources surveys.

Beyond these basic trends, the Shuttle permits a revolution in space design philosophy. The achievement of easy access to and from space makes it possible for the first time to invert a traditional principle of space flight practice, which required the spacecraft to be as simple, light, uncomplicated and reliable as possible, with the complexity of systems and functions relegated to the earth surface: highly specialized, voluminous, immobile, centralized and expensive ground communications equipment, facilities and control centers. With the newly-won capacity of assembly, repair, maintenance and checkout of large space facilities on orbit, we can now go to large, complex, massive and less reliable orbital systems of very long flighttimes, combined with advanced technologies like decentralized microcomputers, multiple-beam antennas, on-board message switching, etc., and commensurately reduce the ground terminals in size, mass, complexity, power usage and costs -- in the case of person-to-person voice communications as much as to the size of a wristwatch. It is clear that only this complexity inversion will make space flight directly beneficial and affordable to users, even for the "man in the street."(7,8,14)

Not only will the philosophy of deliberately making satellites large and highly capable allow the user equipment to be small, highly mobile or portable, and inexpensive, but the vantage point of geosynchronous altitude will also enable one or only a few satellites to service millions of earth-based users and to perform functions not possible with simpler and smaller satellites.

6. Electronic Services from Space

There can be no doubt that in the decade ahead the emphasis in space activities will be on applications satellites, with first priority on these new telecommunications systems. The nations of the Third World, which at present typically are burdened by extremely deficient infrastructures, should be particularly interested in these systems. But in the highly industrialized countries, too, the demand will be immense. Projections are showing this for the United States clearly: the number of telephone calls, in 1950 about 64 billion, including 2.7 billion long-distance calls, by 1970 had risen to 174 billion. For 1990, 482 billion calls are projected(9), including long-distance (10%). Check and credit transactions in 1970 totalled 56 billion -- a number projected to increase *six-fold* within a span of 20 years.

In 1990, there will be in the United States (9):

- 100 billion pieces of first class and air mail
- 1 billion videotelephone conversations (1970: 1.2 million)
- 72,000 hours annual TV viewing
- 20 million remote accesses to libraries for browsing
- 35 million telegrams
- 100 million hotel/motel reservations
- 365,000 pages of newspaper facsimile transmission annually

- 7 million remote patent searches
- 200 million cases of remote medical consultation and diagnosis (9)

Large satellites in geostationary orbit at 35,870 km altitude, equipped with multi-beam high-gain antennas of 220 ft (67 m) diameter, generating 7000 tightly collimated "spot" beams with 230,000 full-time voice channels in the 10 cm microwave range and solar-generated DC power of over 250 kW, could enter operation around that time to handle up to 25 million users of the wrist-radio communications system.(10)(Fig. 6)

A variety of such large public service systems, "ganged" and integrated on a geostationary platform as an "orbital antenna farm"(11) or Public Service Platform (6, 12), will take over a multitude of electronic services like education, medical aid and wide-area health care, electronic mail, news services, three-dimensional teleconferences, telemonitoring and teleoperation, search and rescue, time and frequency dissemination, navigation, electronic commuting, all-nations "hot-lines," railcar and package locator, nuclear fuel tracker, burglar/intrusion alarms, international air/sea traffic control, and many others.(7,13)

Other large complexes in space will serve for data acquisition in the observation of the earth's surface: prospecting for resources, harvest prognostications, food production, water budgeting, pollution alerts and environmental protection, land-use monitoring, weather forecasting and climate research, ocean surveys, atmospheric soundings, tectonic plates dynamics, geological formations studies, mapping and inventorying, etc.

One industry proposal of a public service platform(14) entails a multi-purpose satellite in geosynchronous orbit, weighing 65,000 lbs (30 t) and measuring 788 ft (240 m) in length. Generating 500 kW with solar cells and using a total of 23 antennas measuring from 3.3 ft (1 m) to 60 ft (18 m) in diameter, this system could provide to the continental United States --

- broadcast education over 5 simultaneous video channels for 16 hours a day
- personal voice communications to 45,000 simultaneous users equipped with hand-held communicators
- national information services, comprising instant access to government, university and commercial data banks

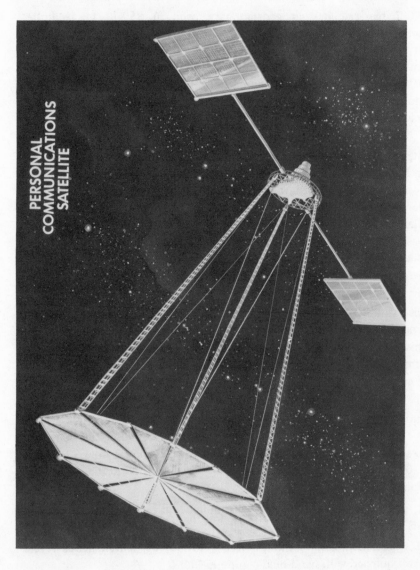

Figure 6 Personal Communication Satellite

• teleconferencing over up to 150 simultaneous 2-way video channels, and

• electronic mail transmission at a rate of 40 million pages per day with overnight delivery from 800 sorting centers

Preliminary studies of other advanced systems(10) have shown that --

• A satellite for educational TV with services to schools, open universities, public TV access and special services, such as medical, vocational and educational aid to migrants and handicapped, could provide 600 spot beams with 634 uplink and 1491 downlink color TV channels, covering 16000 districts and 65000 schools, and would require 63 kW prime power and a 31 ft (9.5 m) antenna.

• A satellite of 6000 lbs total weight, handling all electronic (facsimile) transmission of local and long-distance mail for government and large business corporations, and between the two user groups (1990: 17 billion first class pieces annually), could service over half a million simple terminals with 846 spot beams to as many towns and cities with population of 25,000 or more.

• Direct satellite TV broadcasts of multi-media programming with sign language or subtitles to small rooftop antennas could enrich the lives of 14 million individuals in the U.S. who suffer total loss of hearing, reaching all 48 contiguous states 24 hours a day on channels 74 and 76.

• A teleconference satellite with a 56 ft (17 m) antenna and 220 kW power usage could "bring" remote participants of a conference nation-wide to the place of the meeting as holographic images in three dimensions, color and stereo sound, and do this for 1000 conferences simultaneously, thereby avoiding the disadvantages in energy usage, time, etc., of physical travel.

7. Products "Made in Space"

Next in importance to public services from space is the commercial manufacture of products and goods in orbit. Exploitation of the space environment through processing of commercial inorganic, biological and pharmaceutical materials, as well as through development of new manufacturing methods and processes designed to enhance productivity on earth, is expected to develop extremely high industrial potential. Such products would not only affect world trade and lead to lower costs,

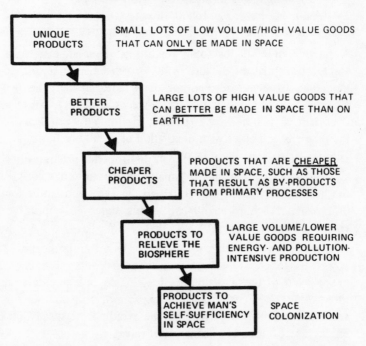

EVOLUTION OF SPACE INDUSTRIALIZATION

EXAMPLE: SPACE PROCESSING & MANUFACTURING OF GOODS

UNIQUE PRODUCTS — SMALL LOTS OF LOW VOLUME/HIGH VALUE GOODS THAT CAN ONLY BE MADE IN SPACE

BETTER PRODUCTS — LARGE LOTS OF HIGH VALUE GOODS THAT CAN BETTER BE MADE IN SPACE THAN ON EARTH

CHEAPER PRODUCTS — PRODUCTS THAT ARE CHEAPER MADE IN SPACE, SUCH AS THOSE THAT RESULT AS BY-PRODUCTS FROM PRIMARY PROCESSES

PRODUCTS TO RELIEVE THE BIOSPHERE — LARGE VOLUME/LOWER VALUE GOODS REQUIRING ENERGY- AND POLLUTION-INTENSIVE PRODUCTION

PRODUCTS TO ACHIEVE MAN'S SELF-SUFFICIENCY IN SPACE — SPACE COLONIZATION

Figure 7   Evolution of Space Industrialization

thus benefiting national and global economy, but they also would be important to human health by contributing to disease prevention and more effective treatment. (Fig. 7)

At present, we know of five basic types of industrial processes that require a zero-gravity environment for improved material quality, more efficient material utilization, commercially significant production volume, and lower production cost. These are:

- *crystal growth* , including growth from melt, in solution, and from a vapor phrase

- *purification and separation* , benefiting from greatly reduced buoyancy and convective effects in zero-g

- *mixing* , particularly applied to immiscible (on earth) materials and composite materials

- *solidification* , including controlled or directionally solidified eutectics, preparation of glasses, and supercooling with homogeneous nucleation

- *fluid processes* , both chemical reactions and physical thermodynamic phenomena

Techniques such as vapor deposition require a high degree of vacuum.

Large, structurally perfect single crystals, obtainable only in the weightless environment, will prove of great value for the electronics industry. Linear extrapolation of the current semiconductor market growth rate, for example, yields a $12.7 billion market in 2000. A conservative estimate would target space-produced semiconductor material as capturing at least 10% of the market, which would grow from about $0.5 million in 1985 to $1.27 billion by the year 2000.(15)

Potential space-made electronic products include compound semiconductors, integrated circuit chips, magnetic switches, relays, magnetic detectors, ultrasonic and optical frequency filters, superconductors, high-power rectifiers, piezoelectric light-emitting diodes, ferroelectrics, inverters, large-area detectors, radiation detectors, holographic storage crystals, and X-ray targets. (16)

Organic products, too, will become important for large-scale industrial production in space, e.g., certain enzymes, vaccines, insecticides, red and white blood cells, nerve cells, hormones such as insulin, erythropoietin, etc. (17)

A prime candidate of high impact on public health appears to be the enzyme urokinase, a catalytic substance produced by specialized cells

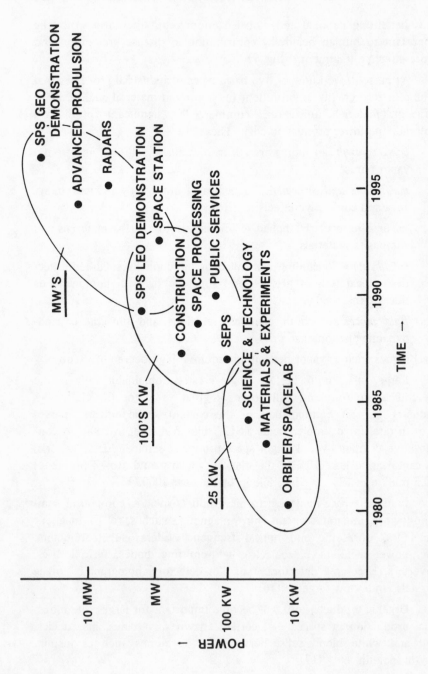

Figure 8   Evolution of Power Requirements in Space

located in the kidneys. Its function is to prevent blood clotting which often causes coronary thrombosis, phlebitis, strokes and pulmonary embolisms. To obtain a single dose of urokinase on earth, costing $1200, about 400 gallons of human urine must be refined. Thus, only experimental quantities are available, but current requirements are for about 500,000 to 600,000 doses/year, to prevent 50,000 deaths/year due to thromboembolisms. In zero-g, kidney cultures can be easily grown. In addition, the specialized cells are producing the enzyme in zero-g at a faster rate than in terrestrial laboratories, as past experiments in space have shown. By applying small electrical forces in a process known as electrophoresis, quantities of urokinase can be separated rapidly with high purity, due to the absence of convection currents and sedimentation in the free-fall condition.(18) Studies indicate that packaged urokinase would cost about one-tenth as much to produce in space as on earth, with the "spacekinase" eventually selling for $75-$100 per dose (including the cost of transportation to and from space).(19)

These products alone are seen to become a multi-billion dollar business in the decades ahead.(20,21,22)

Other applications are expected in the areas of material sciences, metallurgy, composite materials, and creation of new compounds and alloys. High-quality semiconductors, ultra-strong fibers, perfect glasses, large single crystals, high-coercive-strength magnets, which would lead to weight reductions in motors and thus to energy savings -- these are just a few examples of a new, uniquely exciting world of products and goods, at the threshold of which we are -- just barely -- standing.

8. Energy from Space

Even more significant to mankind may be the possibility of one day unlocking the inexhaustible energy source of the sun for our electricity-hungry civilization. According to recent estimates (23,24), the United States electric generating capacity of the next 25 years must grow from about 500,000 megawatt (MW) in 1975 to about 2000 gigawatt (GW) in the year 2000, which amounts to an increase of three times our current capacity. By providing an abundance in space of clean, limitless power, space industrialization could offer a new energy source that would relieve the environmental burden imposed on the biosphere by fossil and nuclear power production. (Fig. 8)

After collection of the power in geosynchronous orbit and photovoltaic or solar-thermal conversion, transmission to earth would be by

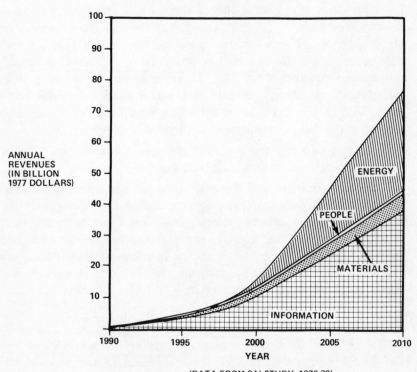

Figure 9   Space Industrialization Projected Revenues

microwave radio beam to a ground receiver antenna for subsequent reconversion to electricity. Such power plants in space will require immense size (typically 50 square kilometers of solar cell arrays for 5000 MW net power available to the ground power grid) and large investments, but the zero-g environment permits constructions that resemble spider webs more than equivalent earth-based structures. By 2000, such energy satellites could provide tens of thousands of megawatts electricity to our energy-starved cities, meeting perhaps up to 20% of the power requirements of the United States at that time.

9. The Humanization of Space

Space industrialization means, first of all, new markets and more jobs here on earth because -- due to its commercial activities and the related service jobs of the "job pyramid" -- it raises the labor-intensive characteristics of the space program more than its capital-intensive features.(25) Thus, it fosters a program that is economically more balanced. (Fig. 9)

But many of the space industries of the decades ahead will also require men and women as highly qualified workers in space, operating from habitable orbital facilities. Various concepts of orbiting platforms and space habitats are under continued study by NASA: man-tended construction systems and permanently manned space stations of modular or "building block" design which will grow by steps, organically -- almost like trees, built up and supported by the versatile Space Shuttle. Along with these studies, plans for a manned reusable Orbital Transfer Vehicle (OTV) which can carry people and cargo to geosynchronous orbit and eventually to the moon, are on the drawing board. Space systems currently envisioned by long-range planners for the 80's and 90's can be seen as stepping stones to and precursors of large earth-orbiting and lunar-based space communities and space settlements of the third millenium, housing hundreds, even thousands of humans.(1,26,27) (Fig. 10)

"Human" industries in space, such as medical, clinical and biogenetic research, space science and space-borne educational centers ("University in Orbit"), space therapeutics with hospital and sanitarium in orbit, and activities in areas like entertainment and the arts are long-range possibilities which will eventually be brought within our grasp through the step-wise development of space by the preceding phases of space industrialization and the further reduction of transportation costs

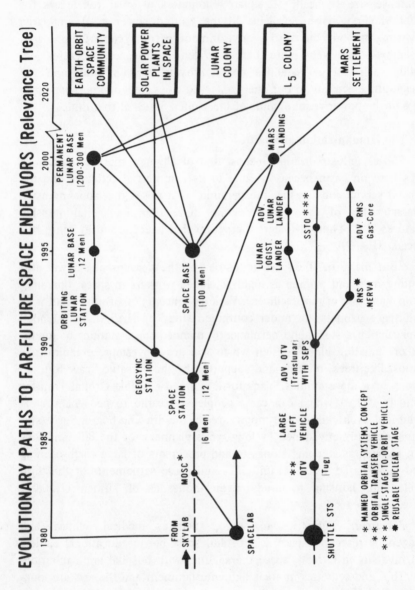

EVOLUTIONARY PATHS TO FAR-FUTURE SPACE ENDEAVORS (Relevance Tree)

* MANNED ORBITAL SYSTEMS CONCEPT
** ORBITAL TRANSFER VEHICLE
*** SINGLE-STAGE-TO-ORBIT VEHICLE
● REUSABLE NUCLEAR STAGE

Figure 10 Evolutionary Paths to Far-Future Space Endeavors

that they will induce. Orbital tourism may certainly become important when people, who today are gladly paying $2000 for a supersonic return flight between Europe and South America or who treat themselves to a $10,000 all-expense cruise around the world, will want personally to experience space and thereby add a whole new dimension to tourism.

From our long-range studies it is clear that the utilization of near-earth space can take on different forms of appearance, to be sure, but several basic requirements are apparent as common denominators for all of them, sure to become characteristic of future space activities and therefore requiring dedicated research and development attention during the early and mid-80's:

- *large* structures in space
- *complex* orbital facility systems
- *high* on-board power requirements
- *long* flight durations (up to 30 years)

These commonality features are interlinked; in a most fundamental sense, they require the presence of humans in orbit for assembly, checkout, servicing, repair and maintenance.

In conclusion, it is clearly evident from today's viewpoint that with the first operational flights of the Space Shuttle in 1980 we are entering a new era of space activities (some, no doubt, will be dragged kicking and screaming into it), in which on the one hand -- and in the near term -- space will be brought down to earth in the service of humans, and on the other hand -- in the longer range -- humans in their natural growth will be led into space: the dichotomy of the *Humanization of Space.*

REFERENCES

1. von Puttkamer, Jesco, "Developing Space Occupancy: Perspectives on NASA Future Space Programme Planning." *J. Brit. Interplanetary Society* , vol. 29, March 1976, pp. 147-173.

2. von Puttkamer, Jesco, "The Next 25 Years: Industrialization of Space -- Rationale for Planning." *J. Brit. Interplanetary Society* , vol. 30, no. 7, July 1977.

3. *Space Transportation System -- User Handbook.* National Aeronautics and Space Administration, June 1977.

4. Disher, John H., "Next Steps in Space Transportation and Operations." AIAA *Astronautics & Aeronautics* , vol. 16, no. 1, January 1978, pp. 22-30.

5. Edelson, B.I., "Global Satellite Communications." *Scientific American* , February 1977.

6. "Federal Research and Development for Satellite Communications." Committee on Satellite Communications (W. B. Davenport, Chmn.), National Academy of Sciences, Washington, D.C. 1977.

7. Bekey, I., Mayer, H. L., and Wolfe, M. G., "Advanced Space System Concepts and Their Orbital Support Needs (1980-2000)." Aerospace Corp. report ATR-76(7365)-1, vols. 1-4, contract NASW-2727, April 1976.

8. Bekey, I., and Mayer, H.L., "1980-2000: Raising Our Sights for Advance Space Systems." AIAA *Astronautics & Aeronautics* , July-August 1976.

9. Hough, Roger W., "Future Data Traffic Volume." *Computer* , September/October 1977.

10. Bekey, I., "Preliminary Definitions and Evaluation of Advanced Space Concepts," Aerospace Corporation, NASA contract NASW-3030, 1977-78, ATR-78(7674),Vols. 1&2, 30 June 1978.

11. Edelson, B.I., and Morgan, W.L. "Orbital Antenna Farms." AIAA *Astronautics & Aeronautics* , September 1977.

12. "A Public Service Communications Satellite." NASA Goddard Space Flight Center, PSCS User Brochure, February 1977.

13. Cooper, Robert S., and Redisch, W.N., "Satellite Communication for Public Services." Paper presented at AAS/AIAA Bicentennial Space Symposium, 6-8 October 1976, Washington, D.C., (NASA Goddard Space Flight Center).

14. "Space Industrialization Study," Rockwell International Corp., NASA contract NAS8-32198, 1976-78, 4 Vols.: NASA CR 150720-150723 (1978).

15. "Space Industrialization Study," Science Applications, Inc., NASA contract NAS8-32197, 1976-78, 4 Vols: SAI-79-602-HU through SAI-79-605-HU, April 1978.

16. Waltz, Donald M., and Hammel, R.L., "Space Factories." Paper presented at the IAF XXVIII Congress, Prague, 25 September-1 October 1977. Preprint No. 77-56.

17. Bannister, T.C., "Materials Processing in Space." Appendix to "Outlook for Space" report, NASA SP-386, January 1976.

18. Barlow, G., "Space Processing of Biological Materials." Paper presented at AIAA/NASA Symposium on Space Industrialization, 26-27 May 1976, NASA/MSFC, Huntsville, Ala.

19. Geschwind, Gary. "The Space Factory." *Horizon* Grumman Aerospace Co., vol. 13, no. 1, 1977.

20. "Industries in Space to Benefit Mankind - A View Over the Next 30 Years." Rockwell International Co. Space Division, brochure SD 77-AP-0094, 1977.

21. "Opportunities in Space Industrialization - The Growing Commercial Use of Space." Science Applications, Inc. brochure, 1978.

22. "Planning for Materials Processing in Space." Final Report of the NASA/ASEE Engineering Systems Design Summer Study at NASA/MSFC, Huntsville, Ala. Grant #NGT 01-002-095, September 1975.

23. Friedlander, G.D., in IEEE *Spectrum* vol. 12, p. 32, May 1975.

24. Statistical Yearbook of the Electricity Utility Industry for 1973, p. 5, table 1-S, Edison Electrical Institute, New York, 1973.

25. Ehricke, Krafft A., "Space Industrial Productivity, New Options for the Future." U.S. House of Representatives Hearings on Future Space Programs, July 1975.

26. Brown, William M., and Kahn, Herman, "Long-Term Prospects for Developments in Space (A Scenario Approach)." Final report, NASW-2924, Hudson Institute, Inc., HI-2638-RR, October 30, 1977.

27. Johnson, R.D., and Holbrow, Ch. (Eds.), "Space Settlements - A Design Study." NASA SP-413, 1977. Government Printing Office, Washington, D.C.

2

Limits to Growth Implications of Space Settlements

Brian O'Leary

Abstract

Engineering studies of G. K. O'Neill's proposal for building satellite power stations from lunar and astroidal materials suggest the cost-effective delivery of large quantities of electricity on Earth by the turn of the century. The retrieval of near-Earth asteroidal resources also makes possible the direct recovery on Earth of large quantities of iron and nickel, possibly at competitive prices. If the concept of closed agricultural stations can be experimentally verified, food could be grown in space-manufactured facilities for terrestrial recovery and consumption. A preliminary cost estimate indicates that the projected world food supply for the year 2000 could be grown in space for an investment on the order of $100 to $200 billion. As soon as thirty years from now, it may become possible to provide extensive, reliable and competitive sources of energy, resources and food for an increasingly impoverished planet, and thus expand the limits to growth. The implications of this program are discussed.

32   Brian O'Leary

1. Introduction

O'Neill(1) has proposed that self-sufficient human settlements could be constructed in space from materials retrieved from the Moon and asteroids. Further studies(2) have confirmed that the technology is available, that the cost would be comparable to that of the Apollo program, and that the first permanent settlement for 10,000 people could be established as early as the 1990's.

In another study, O'Neill(3) suggested that base-load central station electricity could be economically obtained on the Earth by microwave link from satellite solar power stations constructed from nonterrestrial materials. By early in the twenty-first century, these stations could replace fossil fuel and nuclear power plants as a major source of the world's energy supply, at lower environmental cost, with an investment many times less than that planned for the capital expansion of electricity generating equipment on the Earth. The basic feasibility and cost competitiveness of this scheme have also been confirmed by detailed engineering studies, in which a step-by-step Apollo-scale program would be required.(2,4)

These studies have considered the Moon as the source of materials. The cost of retrieval of these materials and transport to a high orbital manufacturing facility has been estimated to be on the order to $1 to $2 per kilogram amortized over 10-20 years.(2,4) This compares favorably with $1000 per kilogram for Earth-launch of materials aboard the Space Shuttle planned for the 1980's and with $100 per kilogram for Earth-launch aboard an advanced heavy lift vehicle which could be developed for the 1990's.

2. Asteroid Retrieval

Earth-approaching asteroids have also been suggested as competitive materials for space manufacturing.(5) The resource is vast; probably more than 100,000 such objects exist with diameters greater than 100 meters (10^6 tons). There are opportunities where the total energy per unit mass for the transfer of some of these asteroids to a manufacturing site in high Earth orbit is comparable to that of lunar materials. The cost may be many times less for logistical reasons: no soft landings would be required, and solar energy for processing and propulsion is continuously available at the asteroid. Optical studies suggest ordinary and carbonaceous meteorite compositions for these asteroids, with some containing large quantities of metallic iron and nickel, and others, carbon, nitrogen and hydrogen, which appear to be lacking on the Moon. Discoveries of several new candidate asteroids over the next few years will allow for a selection of an abundance of materials and mission possibilities. The asteroid-retrieval scenarios have also been subjected to engineering and parametric studies, which have confirmed earlier suggestions that the retrieval cost could become less than $1 per kilogram.(6)

A major conclusion of the study was that, through the techniques of multiple gravity assists by the Earth, Moon and Venus, the total one-way velocity increment (Δv) from Earth escape (or capture) to rendezvous (departure from) with existing asteroids, in favorable cases, is in the range 2 to 6 km/sec. For example, one opportunity was studied in which the retrieval of material from asteroid 1977 HB to high Earth orbit was 3 km/sec. Analysis of the most probable mission opportunities indicates that a single-stage low Earth orbit -to- asteroid mass driver (a solar-powered electric linear motor which can use any available material as propellant) with an exhaust velocity of ~8 km/sec may be the most cost-effective alternative, and is less sensitive to Δv than multiple stage devices with lower exhaust velocities. Further refinements of mission analysis techniques for known objects will permit a more precise determination of opportunities. In addition, opportunities will arise as the number of known objects increases appreciably over the coming years.

The total asteroid-retrieving mass consists of mass driver, mining equipment, and reaction mass for the boost. In the range of likely Δv's to be encountered, the total mass required for launch to low Earth orbit for the retrieval of ~50 per cent of a 1 million ton asteroid is close to 10,000 tons, similar to the lunar case. This magnitude of mass could be launched by an upgraded (Class II) shuttle during the mid-to-late 1980's with the external tanks used as reaction mass, at a cost of approximately $2 billion. Development costs (~$10 billion) would dominate and would be absorbed in subsequent, larger scale asteroidal retrievals.

Perhaps the most unusual aspect of the retrieval of asteroidal materials is the availability of large quantities of volatiles (water and carbon compounds) and free metals. These materials may not be available in large quantities from the Moon. About 30,000 tons of consumables are needed for space settlements to support the construction of satellite power stations. The bulk of these could come from the asteroids rather from the Earth. Moreover, separating free metals at the asteroid and alloying them at the space manufacturing facility (SMF) into useful structures may eliminate many of the complex steps of chemical processing required in the lunar case. Hydrogen and oxygen from a water-rich asteroid could also provide an alternative rocket fuel for the retrieval of materials if structural instabilities prove to limit the effectiveness of mass driver reaction engines in space.

In the scenario which was considered, a significant fraction of the ∼500,000 tons of asteroidal material which would arrive at the SMF could be water and carbon if a type I or II carbonaceous object at a low to moderate Δv could be discovered. The possibility exists that the Earth-approaching asteroid Betulia is carbonaceous . Statistical considerations suggest that such a discovery, in an expanded search and follow-up program, is probable at some time over the next few years. Such an object would obviously be a prime target for a precursor mission.

Further in the future, asteroidal resources could be obtained in even larger quantities. A multiple-phase program could be carried out. The first phase would include launching a mass driver asteroid retrieval mission. The first phase would also include developing space manufacturing facilities sized to process up to 10^7 tons of asteroidal material into satellite solar power plants over a period of about 10 years. In the second phase one such plant would power a mass-driver supertug (∼10,000 Mw, 10^5 tons) to a 1 km diameter (10^9 ton) metal-rich asteroid. This asteroid could provide more than enough material for the world's energy supply from space solar power stations, and sizably increase the availability of iron and nickel on the Earth. Later retrievals of 10^{10} -10^{11} ton asteroids could increase worldwide supplies of food and metal resources. Space habitats, large telescope arrays and planetary explorers could also be manufactured from asteroids in the 10^9 ton-or-more category, possibly by the turn of the century.

It is not the purpose of this paper to review in detail the content of these engineering studies. The basic point is to indicate that any one, two or three of the most fundamental limitations of the Earth's biosphere -- energy, food and raw materials -- could be supplied from near-Earth asteroidal resources in combination with solar energy available in space. One begins to approach a range of cost-competitiveness if the total retrieval cost can be brought down to some tens of cents per kilogram.

The literature on the construction of satellite power stations from nonterrestrial materials makes clear the potential economy of this source of electricity on Earth(3,4) and will not be reviewed here. Although this application of space manufacturing appears to be the most immediately compelling rationale, further tests and demonstrations will be required to assure us that there are no show-stoppers or better alternatives to satellite power. The two other potentials of asteroidal resources for Earth use -- raw materials and food -- have not been treated as widely and are the subjects of the next two sections of this paper.

3. Asteroidal Materials for Use on Earth

Gaffey and McCord(7) have discussed the possibility of retrieving asteroidal materials for use directly as metal resources on the Earth. They suggested the development of vacuum-foam reentry bodies surrounding the asteroidal fragments which would make possible intact landings in the ocean with subsequent towing ashore by barge. With about 10^{10} tons of a metal-rich asteroid landed on the Earth, they projected the market price of iron to be $0.40/kg and nickel, $0.80/kg. Therefore, with the mission energy requirement described here, the cost of transfer by mass driver of an iron-nickel asteroid to the Earth's surface may become immediately competitive. It is interesting to note that the Sudbury Astroblem in Ontario, Canada, was the source of approximately one-half the total world production of nickel between 1961 and 1965.(8) The metal in this deposit is probably meteoric.(9) It is likely that we have already (unwittingly) been mining the asteroids!

4. Food from Space

The possibility of growing food in space for consumption on the Earth,(10) to my knowledge, has not been previously considered. O'Neill(11) has discussed the concept of using the methods of intensive agriculture for growing food in space for the inhabitants of space settlements. He suggests that corn, cereal, bread, poultry and pork could be raised in closed agricultural areas adjacent to the space settlements under controlled conditions where light, temperature and moisture can be varied according to the needs of a given crop. The passage of seasons, droughts, pests and pervasive disease could be eliminated. Fertilizer could be produced from solar process heat in space from nitrogen and oxygen in the asteroids. With an exponentially growing supply of materials available for constructing agricultural facilities, it is possible to build vast areas for growing in space. Although the concept of closed agricultural ecologies has not yet been experimentally verified, it appears likely that, sooner or later, food will be grown successfully in space.

The favorable economics of asteroid retrieval combined with the apparent attractiveness of space agronomy raises the question of whether food for consumption on Earth could be grown more economically and reliably in space than on Earth in a highly developed program of space manufacturing. Large quantities of dehydrated mature crops could be deorbited (possibly by a mass driver device) , enter the Earth's atmosphere aboard a vacuum-foam reentry body, land in the ocean offshore from potential consumers, and be towed ashore by barge for end use.

O'Neill(11) estimated that food grown on an area of 0.008 hectare in space would comfortably feed one person. For a projected world population of 6 billion in the year 2000, the growing requirement would be $\sim$500,000 km^2 growing area. Assuming that the soil and structural mass of an agricultural pod is $\sim$10 cm thick, with most of this mass in the form of water harvested from the asteroids, we would need for the mass of asteroidal material

$$(10g/cm^2)(500,000km^2) = 5(10^{10}) tons.$$

This amount of material could come from a 3 km diameter carbonaceous asteroid. Such an asteroid could be towed in by a mass driver powered by the equivalent of $\sim$100 (or less) 10,000 megawatt satellite solar power stations. (Most of the mass driver expense would go into the powerplant and its radiators). Estimates of the cost of the early phases of space manufacturing(2,4) and the economics of asteroid retrieval(6) suggest that this number of satellite solar power stations could be built at a cost of some hundreds of billions of dollars spread over the next twenty to thirty years. This investment is comparable to the $700 billion capital which Revelle(12) estimates will be required between now and the year 2000 for irrigation development and agricultural modernization in the Third world.

As in the case for electricity, the worldwide market for food is large and is growing. Food production in the United States alone totals approximately $100 billion per year with 20% of this exported. The case for growing food in space may become even more compelling as the food supply for a growing population dwindles, as water resources become scarcer in some growing areas, as energy requirements for agriculture increase, as the use of fertilizers creates greater adverse environmental effects (including potential ozone depletion in the stratosphere), as drought and severe climates continue to cause unwanted fluctuation in food production, and as much of the world continues to hunger.

Although they are clearly preliminary, these estimates seem to indicate that a virtually unlimited and reliable supply of food can be grown for the Earth as early as thirty years from now, at an economically competitive cost and independently of the Earth's biosphere. Such a possibility would relieve the immense pressures the human race now feels in managing its limited supply of nonrenewable energy, food and other resources. The onset of space agriculture may occur just in time to avert massive starvation.

5. Implications for Limits to Growth

The potential impact of using resources available in near-Earth space on current debates about growth on Earth is enormous. Until recently the assumption was implicit that humanity must depend on the Earth's biosphere for survival. The growth/no-growth debate can be summarized as follows:

Growth advocates feel that a rapid transition to a steady state is not possible with the existence of poorer nations seeking to raise their standard of living to that of the western nations.(13) The transition itself, they argue, is contrary to the tenets of existing political systems and would appear to be impossible, on purely practical grounds for at least decades. They suggest that a steady-state world would impose severe restrictions on individual freedom and opportunity.

The steady state advocates, on the other hand, feel that growth must stop for human survival.(14) With the limits to growth models taken at face value, even allowing for some room for guessing exactly when they will be reached, the no-growth argument is that political systems and lifestyles must adapt to these limits or suffer the consequences which any biological system must when it overbreeds and runs out of prey. There seems to be little dispute that this reasoning is valid when applied on a global scale.

Others have attempted to reconcile these views by acknowledging the seriousness of the problem and proposing long-term Earth-based solutions. There is, for example, Benoit's dynamic equilibrium economy(15) which advocates a shift in growth from the depletion of non-renewable resources to the rapid development of renewable resources, elimination of waste, reduction and eventual stabilization of population growth, and reliance on science and technology to provide innovations for increased quality of life. Forrester (16) has argued that more attention needs to be paid to social rather than physical limits. Solutions should be on a national scale and in a time frame of decades. Shumacher(17) has proposed that technologies and human activities will need to be developed on a small scale, with accountability by individuals and small groups rather than the amorphous large organizations which currently control our destiny. Whether these measures are politically possible or would be adequate to turn the tide is open to debate; there appears to be skepticism that these proposed solutions could be realistically implemented before widespread disaster strikes.

The space solution to the limits to growth transcends the dilemma -- at least on a time scale of decades to centuries. The terrestrial limits to growth of food, energy and material resources, of pollution, and of population, could be relieved in a matter of years after the first retrieval of materials from the shallow gravity wells of the Moon or asteroids.

Eventually large scale industry could move into space, relieving the Earth of pollution pressures. Population stresses could be further relieved by the option to live in space. The space solution provides new outlets for the further development of human exploration (first the solar system, later the stars), basic science, social experimentation in a variety of settings, and an increased potential for human survival. Some writers have suggested that we are on the verge of a major new step of evolution, analogous to the first animal movement from water to land.

A growing number of scientists and writers feel that such a project is an obvious and perhaps inevitable solution to the limits to growth on Earth.(11,18) I share that view, with the caution that the implications for controlled exploitation and economic gain could further upset the social order and create new difficulties on a massive scale.

No-growth advocates and ecologists, particularly, have pointed to some of these problems.(19) An exploitation of space which merely provides mankind a relief valve from the biosphere could result in a cancerous one-way growth into the solar system and would merely buy time before coming up against new limits. Visions are conjured of profit-hungry exploitation, creating energy overdemand and the relaxation of dealing with our problems here on Earth now that we have a panacea.

In other words, the solution to a critical human problem-- shifting consumption from non-renewable resources on Earth to renewable and non- terrestrial resources and raising the standard of living of the poor countries-- may help to create a new problem: exploitation on a larger scale and relaxation of standards. The asteroids contain enough material to provide abundantly for populations tens to thousands of times that of the Earth currently, and solar energy in space is abundant. Civilization may opt for the "Dyson Sphere," a shell of mankind centered around the sun to capture all its energy.

6. Summary

These considerations should provide ample warning that an international political framework for controlling the exploitation of space must be developed. About 50 million tons of material -- an excavation on the Moon one kilometer square and 15 meters deep, or a 300-meter diameter asteroid -- is needed to construct enough power satellites to supply the world with all its energy in 2000. Later retrievals of much larger quantities of asteroidal materials could provide the Earth with abundant food and metal resources. Eventually it may be desirable to establish large colonies for developing exploratory opportunities and alternative lifestyles. In any case, I think that given the possibility of a desperate situation here on Earth in the relative near term, the best hopes of mankind require a vigorous effort in establishing something like Benoit's dynamic equilibrium economy *and* space manufacturing and asteroid retrieval. To hesitate on the basis of vague fears of runaway growth and environmental deterioration, in my opinion, would be unwise and contrary to human nature.

The need to plan for ten, twenty and thirty years from now is essential. The stakes are too high to be tentative in exploring the space manufacturing and asteroid option. Regardless of the outcome of debates on growth, energy policy and political philosophy, the evidence is clear that research and development of space-based manufacturing from non-terrestrial materials should begin immediately. The alternative appears inevitable and undesirable: growing international inequities in energy and food supply, future oil embargoes, expensive energy, depletion of non-renewable resources, proliferation of dangerous nuclear technology, massive hunger and starvation, and tighter controls over human freedom.

Starting a logical progression of steps required to use the vast resources and energy available to us in space demands a unity, discipline and perspective now lacking because of a collective tentativeness in trying something so totally new. The challenge is more one of communication than one of technical feasibility and economics. History makes it clear that this project will go ahead sooner or later. The arguments seem compelling that this is the time to begin to fulfill some of mankind's highest hopes.

"What we need is a Copernican revolution and we don't have it," said Dennis Meadows.(20)

We do have it.

REFERENCES

1. O'Neill, G.K., "The Colonization of Space," *Physics Today* , Vol. 27, No. 9, September 1974, pp. 32-42.

2. O'Neill, G.K., "Engineering a Space Manufacturing Facility," *Astronautics and Aeronautics* , October 1976, pp. 20-28; Driggers, G.W. and Newman, J., "Establishment of a Space Manufacturing Facility," 1976 NASA Ames/OAST Summer Study on Space Manufacturing of Nonterrestrial Materials, *Progress in Astronautics and Aeronautics* , Vol. 57, ed. O'Neill and O'Leary (American Institute of Aeronautics and Astronautics, New York, 1977).

3. O'Neill, G.K., "Space Colonies and Energy Supply to the Earth," *Science* , Vol. 190, December 5, 1975, pp. 943-47.

4. O'Neill, G.K., "The Low (Profile) Road to Space Manufacturing," *Astronautics and Aeronautics* , March 1977, pp. 18-32; Vajk, J. Peter, Engel, Joseph H., and Shettler, John A., "Habitat and Logistic Support Requirements for the Initiation of a Space Manufacturing Enterprise," *The 1977 Ames Summer Study on Space Settlements,* in press; O'Leary, B., "The Construction of Satellite Solar Power Stations from Non-Terrestrial Materials: Feasibility and Economics," Alternative Energy Sources: A National Symposium, Miami, Florida, December 5-7, 1977.

5. O'Leary, B.T., "Mining the Apollo and Amor Asteroids," *Science* , Vol. 129, July 22, 1977, pp. 363-66; O'Leary, B.T., "Mass Driver Retrievals of Earth-Approaching Asteroids," The Third Princeton/AIAA Conference on Space Manufacturing Facilities, American Institute of Aeronautics and Astronautics, 1977.

6. O'Leary, B., Gaffey, Michael J., Ross, David J., and Salkeld, Robert, "Retrieval of Asteroidal Materials," *The 1977 Ames Summer Study on Space Settlements,* in press; Bender, David F., Dunbar, R. Scott, and Ross, David J., "Round-Trip Missions to Low-Delta-V Asteroids and Implications for Material Retrieval," *The 1977 Ames Summer Study on Space Settlements,* in press; Gaffey, Michael J., Helin, Eleanor F., and O'Leary, B., "An Assessment of Near-Earth Asteroid Resources," *The 1977 Ames Summer Study on Space Settlements,* in press; O'Leary, B., "Asteroidal Resources for Space Manufacturing," International Astronautical Federation XXVIIth Congress, Prague, September 25-October 1, 1977; O'Leary, B., "Resource Potentials of Asteroid Capture. Macro-Engineering: The Infra-Structure of Tomorrow," AAAS Annual

Meeting, Washington, D.C., February 13, 1978.

7. Gaffey, M.J. and McCord, T.B., "An Extraterrestrial Source of Natural Resources," *Technology Review* , in press.

8. Boldt, J.R., *The Mining of Nickel: Its Geology, Mining and Extractive Metallurgy* , Methuen & Co, Ltd., London, England, 1967.

9. Dietz, R.S., "Sudbury Astroblem, Splash Emplaced Sub-layer and Possible Cosmogenic Ores," in *Geol. Assoc. of Canada, Special Paper No. 10* , 1972, pp. 29-40.

10. O'Leary, B.T., "Food and Raw Material Supply from Space to the Earth," Seminar on World Scarcity of Food and Raw Materials, Santiago, Chile, January 1977.

11. O'Neill, G.K., *The High Frontier: Human Colonies in Space* , Wm. Morrow and Co., New York, 1977, p. 65.

12. Revelle, R., "The Resources Available for Agriculture," *Scientific American* , September 1976, p. 165.

13. Heilbroner, R., *An Inquiry Into the Human Prospect* , W. W. Norton, New York, 1974; Rustin, B., "No Growth Has to Mean Less is Less," *The New York Times Magazine* May 2, 1976.

14. *e.g.* , Daly, H.E., *Toward a Steady-State Economy* , H.H. Freeman and Co., San Francisco, 1973; Meadows, D.H., *et. al., The Limits to Growth* , Universe Books, New York, 1972.

15. Benoit, E., "The Coming Age of Shortages," *Bulletin of Atomic Scientists* , January, February, and March 1956.

16. Forrester, J.W., Limits to Growth '75 Conference, Houston, Texas, October 20, 1976.

17. Shumacher, E.F., *Small is Beautiful* , Perennial Library, New York, 1975.

18. Heppenheimer, T.A., *Colonies in Space* , Stackpole Books, Harrisburg, Pa., 1977; Hubbard, B.M., *The Hunger of Eve* , Stackpole Books, Harrisburg, Pa., 1977; Feinberg, G., *Consequences of Growth: The Prospects for a Limitless Future* , Seabury Press, New York, 1977; Leary, T., *Neuropolitics* , Starseed Peace Press, Los Angeles, 1977; Stine, G.H., *The Third Industrial Reveolution* , G.P. Putnam Sons, New York, 1975; Vajk, P.J., "The Impact of Space Colonization on World Dynamics," Lawrence Livermore Laboratory UCRL-77584, November 14, 1975.

19. *Co-Evolution Quarterly* , Spring, 1976.

20. Meadows, D.H., Limits to Growth '75 Conference, Houston, Texas, October 20, 1975.

3

Space Exploration: Prospects and Problems for Today and the Future

Leonard W. David

Abstract

This paper argues that limits to our ability to expand in space are placed by warfare extended into space, conflict over resources in space, and pollution of space.

The 1967 space treaty provided for the freedom of exploration, freedom from claims of sovereignty, and freedom from nuclear weapons in space. The terms of this treaty have become outmoded by advancing technology, and the large number of military satellites. Current U.S. Air Force doctrine expresses the need to insure that no other nation gains a strategic military advantage through exploiting the space environment. The Soviet Union has indicated they view the space shuttle as a killer satellite.

The position advocated by many around the globe, that resources need to be viewed as a common heritage of mankind, is a clear precedent for space resources. There is as yet no viable model for international resource development (as opposed to research, such as in Antarctica). Some equatorial nations have advanced claims of sovereignty over the geosynchronous orbital stations above their countries. There is no institutional apparatus for the settlement of rival claims in space.

Without specific action to prevent it, the year 2000 could see the presence of a Saturn-like ring of space garbage around the earth. Planning for space activities has so far been deficient in not calculating its environmental impact.

The promise of space colonization and societal reaction is discussed along with the possibility of planetary engineering -- making new worlds for our species.

The paper concludes that space involvement can offer great promise for the future of mankind, or provide a new divisive element of societal growth.

-editor

1. Introduction

By way of space exploration we have briefly sailed the cosmic ocean and, in the process, found that we will not fall off its edge. Space exploration and application is now at an evolutionary/revolutionary turning point, a mere 20 years old and producing an explosion of scientific inquiry and achievement.

In our conduct of space exploration we have opened to question the future of our species -- earthkind. The apocalyptic "limits to growth" scenario, a planet succumbed to pollution, out-of-control population, limited food and energy resources, and international and communal conflict, is directly challenged by the offerings of space involvement. The consideration of the open-ended option of space utilization reduces the fatalism of limits to growth to what it truly represents -- a limit on creativity and imagination.

However, along with this promise, do we have the common sense to learn from our earth-bound history? Or, will we merely perpetuate into space a myriad of problems we had wished to escape? What is needed is the creation of an orderly movement into space, balanced with social, environmental, political, legal, economic, as well as educational responsibilities. No longer can space exploration be viewed as a technical exercise, but a venture which calls upon all disciplines to assist in its development, sustainment, and productivity.

The carte blanche funding days of the 1960's have been altered, subject to the dynamics of political, social and economic forces. Our next phase of exploration is developing from retrospection and a sorting out of viable and reasonable options. It is in the creation of this next phase of space activity which provides a unique opportunity to perpetuate both the best and the worst traits of our species.

Motivations behind space involvement are mixtures of national prestige, military superiority, scientific exploration and commercial exploitation. It is the manipulation of these factors that will produce degrees of human involvement ranging from peace to war. International space stations, global information systems, satellite solar power stations, space colonization, interstellar flight, and contact with other "intelligent" life forms, are but a few of the opportunities that lie ahead. But in the effort to "humanize" space, we must assure that we do not plant the seeds for earthkind's very destruction. It is a time of cautious optimism and justified hope.

2. War in Space

The possibility of armed conflict in space will do much to inhibit earthkind's expansion -- to break the bonds of limits to growth. The fundamental legal instrument in the field of space law is the 1967 space treaty on *Principles Governing the Activities of States in the Exploration and Uses of Outer Space, including the Moon and Other Celestial Bodies.* This agreement provides that outer space and celestial bodies are freely available for exploration and use by every State and cannot be subjected to any claim of national sovereignty or exclusive use. The Treaty relates that no State may orbit nuclear weapons or install them on a celestial body, and calls for international cooperation in space as a fundamental objective for the community of nations. Lastly, the Treaty declares that space activities should be conducted with a view to benefiting all mankind, however disparate may be their earth-based politics or economic situation.(1)

It is clear, however, that this Treaty is having its growing pains, particularly in keeping pace with what could be called the "militarization" of outer space.(2,3) As questioned by President John Kennedy, is space becoming "a sea of peace or a new terrifying theatre of war?"(4) Today we find in space an entire stable of military communication, weather, navigation, and reconnaissance satellites, a majority of which are under the control of the two superpowers, the United States and the Soviet Union. The number of countries which will become involved in space or space-related activities in the future appears to be growing at a steady rate.(5) As orbital activity increases so has dependence on the "space segment" to provide services for both military and civil utilization. In the future such dependence may include large industrial manufacturing plants and possible energy supplementation from satellite solar power stations (SSPS).(6)

Current justification of the U.S. military involvement in the space medium is firmly established in logic, economics, and the assigned objectives of protecting the national security. These justifications are:(7)

- Uniqueness - some functions essentially can only be done from space, such as near real-time warning of a ballistic missile attack

- Economics - some functions are more cheaply done from space, such as long haul communications

- Function effectiveness - some functions are more effectively done from space, such as meteorology

- Force effectiveness enhancement - some space functions can greatly enhance the effectiveness of terrestrial forces such as battlefield surveillance.

A prime difficulty in establishing distinctions between the peaceful and military use of outer space is that any distinction is arbitrary. The goals of national prestige, military superiority, scientific exploration and commercial exploitation, and their interdependence, override clear distinction. The term "national security" provides a common thread between these indistinguishable space objectives.

As stated in Air Force Manual 1-1, U.S. Air Force Basic Doctrine, January 15, 1975:

> The underlying goal of the U.S. national space policy is that the medium of space must be preserved for peaceful use for the benefit of all mankind. Air Force principles relating to space operations are consistent with this national commitment. National Policy and international treaties restrict the use of space for employment of weapons of mass destruction. *There is, however, a need to insure that no other nation gains a strategic military advantage through the exploitation of the space environment.* (my emphasis)

3. Fear of Military Dominance

The fear of a technological, space-originated "Pearl Harbor" is evident throughout public and military literature. In a Fall, 1974 issue of *Strategic Review* we find that

> despite wishful thinking to the contrary, man is and promises to remain an aggressive, combative creature. We fear, we hate, we fight one another. Until we remove causes of fear and hatred and correct the conditions which prompt us to arm ourselves, we have no choice but to prepare to defend ourselves against attack in whatever form and through whatever media attacks may come. Today and henceforth the United States must be prepared to defend itself against aggression *in* space and *from* space. We cannot surrender the "high ground" without contest.(8)

Still others are openly calling for the development of capabilities which ensure the "defense of the freedom of passage in space."(9)

As noted by space lawyer Andrew Haley:

...a nation is justified in protecting itself from attack no matter where the staging area of the attack may be, including on the high seas or in outer space, and a nation may carry its defensive forces to such areas. The great unresolved problem, so far as defensive measures in space are concerned, is to translate the general recognition of this right of self-defense into some workable criteria for distinguishing between the defensive and offensive uses of space.(10)

As of yet, there appears no clear distinction of what connotes "peaceful" uses of space. Peaceful under one interpretation, primarily used by the United States, means "nonaggressive." A second interpretation of the terms means "non-military" and is used, although not exclusively, by the Soviet Union. In addition, "peaceful" uses has been applied to defense support space missions which are "non-interfering" or "non-aggressive."(11)

4. The Technology of War in Space

A review of current outer space agreements is needed, particularly in light of new space technologies and arms control strategies. It is obvious that technology is continually ahead of the vocabularies of existing space treaties. New technologies or new tactical philosophies develop new vocabularies, voiding past legal commitments. A case in point is recent warning from Pentagon and intelligence community officials, alarmed by "rampaging technology":

Sources say lasers and a range of other high-technology weapons being developed would make arms limitation talks "virtually impossible" by precluding any feasible way to check compliance. In addition, they say, the advanced weapons would defy conventional forms of strategic defense planning.(12)

With the enhancement of military operations by space-based systems, comes fear that an aggressor nation could "deny access to such services." U.S. military officials now warn that U.S. dependence on space systems may have created the proverbial "Achilles heel." This fear is based upon recent Soviet testing of anti-satellites, designed to intercept a satellite, explode near it, and shower the target with fragments. Although the U.S. and the Soviet Union are attempting to stop operational deployments of such weaponry, the U.S. has initiated plans to develop its own anti-satellite.(13) Both Soviet and American systems adhere to the current outer space treaties by being non-nuclear and capable of limited crippling capability.

Coupled with the U.S. decision to develop an anti-satellite capability will be the possible programming of U.S. satellites to deter and even withstand hostile attack. Future military satellites will include the ability to conduct special maneuvers in orbit, to avert anti-satellite assault. Some satellites could be designed to use pulverized aluminum particles to form a "smokescreen," protecting them from destructive laser beams. Also contemplated will be the placing of warning sensors on U.,S. satellites which would transmit signals at first indications of tampering. Other military spacecraft will carry internal radioisotope electric power generators to eliminate vulnerable solar panels.

5. Military Man in Space

With the introduction of the jointly developed NASA/Department of Defense Space Shuttle, military occupation of space is assured. Recent reports indicate that the Soviet Union as well is developing a similar spacecraft.(14) Lt. General Thomas W. Morgan, Commander of the Space and Missile Systems Organizations, AFSC, has stated in the past that the Space Shuttle

> ...will open a new chapter of our national space program. It may well make economically feasible for the first time whole new missions in space -- in addition to opening the door to better or cheaper ways of performing traditional missions. I see the 1980's -- the time in which the STS (Space Transportation System) becomes a proved quantity -- as a time of major reappraisal of the role of space in the Air Force future.(15)

The ability of the Shuttle to transport quantities of men and equipment into space, leading to the construction of large military platforms, has led some to call for development of a U.S. Space Force.(16) It is conceivable that manned military outposts, established in space, could provide real-time strategic information to guide troop movements on the earth below. One could easily imagine the creation of a manned military space industrialization complex, producing goods of strategic value as well.

Due to valuable contributions the Space Shuttle offers to the military, fear of anti-satellite attack, destroying the vehicle, has been considered. (17) Likewise, the Soviet Union has indicated it views the Shuttle as a killer satellite.(18)

Historical analogy with the development of the airplane as a vehicle of war may be useful. Early uses of the airplane as a military weapon were considered to be

> ...merely a reconnaissance tool, designed to be the eye of the ground forces, as balloons had been in earlier wars. But planes were so successful in reconnaissance that it became important for each side to destroy those used by the other. Planes were made with greater speed, and armed with machine guns; thus the fighter aircraft was born.(19)

Will future Shuttle craft be armed, possibly with lasers or charged particle beam weapons?(20)

Policy decisions must soon be made to prevent military domination of space. Military space utilization has developed a dependence on volumes of orbital space to fulfill its assigned tasks. In essence, traditional strategic targets have been moved from the earth into space, requiring protection from aggressive elements. To prevent an aggressor nation from denying access to space-based operations, defense mechanisms will be instituted by the operating country. What will remain is the difficult task of identifying offensive/defensive use of space.

Regardless of agreed-to treaties, both the United States and the Soviet Union are involved in developing a space "game plan." The numerous semantic "technological loopholes" provide the opportunity to consider space as nothing more than an extension of normal arenas of conflict -- air, land and sea. Existing space law was written at a time when our involvement provided no clear picture of the future. Need of distinctions between military, peaceful and aggressive uses of space remains a stumbling block to a "steady-state" of space peace. There exists no international body which has the power to enforce ruling as to violators of signed treaties -- treaties which have been circumvented by new technologies and new vocabularies: i.e., non-nuclear (laser?), limited destruction anti-satellites.

We are now witnessing the justification and development of war strategy which will turn space into a battleground. In the program to humanize space, are we also developing the capability for its very destruction? As we extend earthkind's activity, first to near-earth space and then outward, will military involvement prevent or merely increase the chances of armed conflict?

6. The Natural Resources of Space

After the launch of Sputnik I in 1957, a man in Chicago claimed that he had filed a deed to space with the Cook County (Illinois) recorder in 1949. He charged that the Soviets were trespassing on his space nation of "Celestia" and refused to grant permission for the satellite to cross his nation. Although his claim gathered little publicity and no legal support, the issue of space ownership will play a dominant role in future space progress.

Within the last several years, studies indicate the potential for large scale utilization of the space environment, eventually keyed to mining of the moon and asteroids.(21,22) In addition, exploitation of the sun's energy via Satellite Solar Power Stations (SSPS) is envisioned, along with added use of geostationary orbits for earth-assisting services.(23,24)

Increasingly, exploitation of space for such activities is looked upon with skepticism by developing countries. Particularly in the mining of celestial bodies, arguments have been raised which deserve consideration in the planning of future space activities. Related to the issue of celestial mining and space exploitation are negotiations on using currently inaccessible accumulations of resources on the deep seabed.

7. Seabed Mining - An Analogy

Research indicates quantities of resources are available for mining on the ocean floor. Taking the form of nodules, large amounts of copper, iron, nickel, manganese, and cobalt lie on the ocean bottom. It has been estimated that some 1.5 trillion tons of these nodules exist in the Pacific Ocean basin alone.(25) Although first discovered late in the 1800's, only in recent years has the technology been available to mine the nodules, down to 12,000 feet and below. With the natural ores of earth rapidly being depleted, potential seabed mining has stimulated the establishment of several mining consortia, ready to haul in both resources and profit.

Since 1958, the United Nations has hosted debate between some 150 nations in an attempt to unravel a most complicated question -- who has the right to mine the ocean? Although a number of agreements have been reached, arguments have arisen on such items as: the extent of the territorial sea and the related issues of guarantees of free transit through straits; the degree of control that a coastal state can exercise in an offshore economic zone beyond its territorial waters; and creation of an "international system" for the exploitation of the resources of the deep seabed.

Developing countries, such as Mozambique, Libya, India, and Algeria, are challenging the right to free access to the deep seabed riches. These countries are not technologically equipped to mine the ocean floor and are calling for creation of a Seabed Authority. This Authority would exercise arbitrary power over seabed development and would maintain direct control of all mining operations, using the U.N. General Assembly as a regulatory body. Use of the General Assembly, with its one nation/one vote rule, would bypass the U.N. Security Council and the veto rights of the U.S. and other major powers.

Developed countries, including the U.S., favor a position in which a Seabed Authority would permit qualified countries and private entities, on a non-discriminatory basis, to mine areas of the seabed.

This position is countered by the developing countries who declare that resources of the ocean floor belong to the "Common Heritage of Mankind" (hereafter called CHM). This statement argues that, regardless of who mines the seabed, the resources brought to the surface belong to all mankind. In addition, many coastal countries have extended national boundaries from 12 to 200 miles, labeling it their "economic zone." This labeling protects the area from outside mining interests. Interpretation of the CHM concept has now deadlocked negotiations at the U.N. Law of the Sea Conferences.(26)

8. The Moon, Asteroids, and CHM

In August of 1977, a panel of geologists, chemists and physicists recommended that NASA begin a program of space prospecting, with mining the moon and asteroids as a target by the year 2000.(27) Forming a "Near Earth Resources Workshop," these scientists have urged the National Aeronautics and Space Administration (NASA) to initiate a geochemical survey of the moon's entire surface, as well as mapping and classifying near-earth asteroids.

Initial studies have shown that perhaps thousands of near-earth asteroids could be available for a space factory operation. These gigantic space "nodules" are rich in amounts of iron, nickel, carbon compounds, and perhaps quantities of water. It is being suggested that asteroids could be hauled in to the factory by a space tug vehicle for potential mining operations.

A preliminary report from the study panel to NASA concluded that a significant level of useful material from the moon and asteroids could be obtained by the year 2000.

As in the seabed debates, the international community, particularly the developing countries, are disputing future operations to mine the wealth of minerals found in space. Since 1971, efforts have been made to create a U.N. moon treaty, establishing that resources found on heavenly bodies are beyond limits of national jurisdiction.

At a recent United Nations session, a representative of Chile stated:

> Our delegation is still convinced that there are no valid reasons for not applying to the natural resources of celestial bodies, and in particular the moon, the same regime of "the common heritage of mankind" which has been established for the seabed beyond the limits of national jurisdiction. Such resources should not be the property only of those who are able to explore and exploit them; rather they should be distributed for the benefit of all mankind without discrimination whatsoever, and States should commit themselves to establishing an international regime and procedure to regulate those activities.(28)

It has been suggested that future moon-mining operations could revolve around an international moon base. Analogies have been drawn between the scientific base at Antarctica and an international moon-mine camp. (29,30) However, this model may be crumbling. As with ocean mining, new technologies permit exploitation of previously inaccessible resources of oil, gas, and millions of tons of fish, located off the shores of Antarctica. Those countries, once held together by a bond of scientific inquiry and cooperation, find themselves staking out the Antarctic wasteland, to capitalize on valuable resources. A 1959 treaty which froze all claims of territoriality for 30 years is now viewed as obsolete, due to the promise of exploiting the Antarctic wealth.(31) Again, the CHM theme plays a dominant role, with third world countries demanding a share of the Antarctic riches. So will it be with mining the moon?

9. Geostationary Real Estate

Geostationary orbits, over 22,000 miles above the earth's equator, allow satellites to remain stationary and view 1/3 of our planet from the vantage point. At that altitude, spacecraft travel at the same rotational speed as the earth; to an observer on the ground, such a satellite appears stationary. Numbers of weather, reconnaissance, and communications satellites, both military and civilian, reside in such orbits.

Equatorial countries are claiming that these orbits, centered above their territory, are a "natural resource" of their nations. Objections are not raised as to satellites which pass through or outside their geostationary orbit.(32)   But for devices such as communication and weather satellites or possible future satellite solar power stations, the equatorial nations are demanding that the operating nation request permission to use the orbit, and then conform to the national law of the country over which the satellite is based.   A sidenote to this argument lies in the fact that the United Nations has yet to define where national territories end and space begins.

An added problem in using geostationary orbit is the necessity to allocate communication frequencies as well as orbital space. Standards have been set by the International Telecommunication Union (ITU), calling for satellites to maintain positions with no more than one degree of drift.   Some electronic interference between satellites has already been reported.   As new orbital telecommunication services develop, new regulations will be required to adequately and fairly carve up the "natural resource" both of the electromagnetic spectrum and of the geostationary orbit.(33)

Future arguments may be occasioned by solar energy collected in space. Development of solar power satellites may only come by way of international cooperation, with energy produced by the satellites available to all countries on earth.   Again the CHM philosophy would come into effect.(34)

10. Problems of International Cooperation

The outcome of the Law of the Sea Conference talks, an eventual international treaty, will set precedents that will apply to both space mining operations and the allocation of other space resources. But such international regimes could prove difficult to establish.

International cooperation and goodwill may be in conflict with national security.   As stated in a recent U.S. House of Representatives report, current guidelines for the export of U.S. technology place international goodwill second, compared to preservation of technological leadership. The report raises these important questions: because of the military potential of space technology, should sharing of such technology come under these guidelines? Would maintenance of U.S. technological leadership in space under these guidelines be compatible with U.S. stated policy regarding international cooperation in space? If maintenance of U.S. technological leadership in space can only be achieved at the expense of international cooperation, which goal shall prevail?(35)

Writer Dandridge Cole once likened the moon to the "Panama Canal" with the riches of a deep space "Pacific" beyond.(36) With the development of space resources, claims of the extent of economic need, ideology, national ambition, and technological capability will prove great hurdles to fair and equal use of space resources. With full scale "nationalistic" exploitation of the moon, asteroids, and free space, are we also creating areas of strategic importance? If international cooperative measures fail, will the exploitation of space add a burr in the side of "peaceful uses of outer space," leading to an extension of territoriality and requiring military protection?

The deep seabed negotiations embody answers to these questions. Failure to reach legal consensus could lead to political turmoil, and unrestrained commercial and military rivalry. Much is to be agreed upon here on earth before we pull up anchor and arrive at our next port of call -- space.

11. Space Pollution

If space is a cosmic ocean, then we have surely polluted its beach front. It is not too early to begin a movement for the environmental protection of space. The accidental re-entry in 1978 of the Soviet Cosmos 954 satellite, complete with nuclear reactor, served as a calling card from a future problem that is receiving little attention -- space pollution.

Now orbiting planet earth, thousands of objects mark our first 20 years of space exploration -- a blend of operating and non-functioning satellites, nose cones, rocket engines, and assorted "space junk. It has been estimated that by the year 2000, satellite debris will have created a Saturn-like ring of space garbage around the earth.(37) This debris ring will be comprised of tons of fragments, varying in size, arising from an overcrowded space population of satellites and related debris. As the satellite debris density increases, collisions would occur producing additional fragments. The result is an exponential increase in the number of objects that in time create a belt of debris. Hyper-velocity orbital collisions might become commonplace, due to increasing space traffic.

The Space Shuttle in the 1980's may alleviate some of this problem, coupled with a space tug and retrieval mechanism. Acting as a garbage collection service, cleanup operations could be undertaken at various altitudes and inclinations of orbit. It may also be necessary to construct space vehicles in such a way as to minimize the associated space debris normally deposited in space.

With only limited operations to date in space, experience has shown that contamination of the space environment is detrimental to earthkind's space activity. This contamination can come through outgassing of metals used to construct spacecraft; stabilization propellant, such as hydrazine; accidental spillage of nuclear material; the dumping of human waste products into free space; or frequency pollution caused by communications systems. In the past, some spacecraft have developed a surrounding atmosphere of contaminates, hampering scientific experimentation.(38)

It has been suggested that dumping waste heat energy from industrial space centers into space would not interrupt the "normal heat flow of the universe." Likewise, space industry garbage disposal could be facilitated by locating the refuse on celestial bodies or flinging it into space, never to be seen again.(39) Such sweeping generalities and solutions should be subject to some scrutiny.

Plans for celestial resource mining must include measures of environmental protection. Current information is scarce regarding the ecosystem of space. Some boldly state that future asteroid mining will cause no great concern for the "long term ecological integrity of the solar system."(40) Possible use of non-terrestrial material, in pelletized form, as an "in-space" propulsion method is being evaluated.(41) This concept would create "zones" where the discharged particles would travel. The result would be "no trespass," "caution," and "in the green" areas, dividing space into regions of safe and not-so-safe transit.(42) These views should be seriously questioned.

The environmental protection of certain areas of earth represents an attitude that should be reflected in space exploitation. On earth, we find environmental concern regarding the ecological effects of mining and drilling in such wastelands as Antarctica and Alaska. With the world's weather largely effected by Antarctica, scientists are expressing concern as to possible environmental consequences of large scale exploitation.(43)

A similar concern has led to more than half a century of debate regarding utilization of the Alaskan wilderness. Arguments center on use of Alaska's timber, wildlife, oil and minerals. New legislation will possibly lead to designations of large Alaskan areas as "wilderness areas," a classification which generally prohibits commercial development.(44)

Will future space environmentalists demand similar restrictions on moon and asteroid mining or require environmental impact statements for space industrialization?

Could our starry night become punctuated with large "spacescrapers," or dotted with solar power satellites, strung like Christmas lights? Would this constitute a form of visual pollution -- billboards of human activity hung in space? Perhaps such activity will become commonplace -- similar to oil drilling rigs, stationed off our coastlines: markers of human technology and progress?

Space pollution control must be inherent in the growth of earthkind's orbital capabilities. Use of the earth's oceans as a dumping ground for everything from human sewage to radioactive material should have taught us important lessons. The philosophy of an "out of sight, out of mind" solution to pollution has proven to come back and haunt us. True, space is a new vast ocean, but only in relative terms of travel speed and distance. New forms of propulsion, which will most assuredly come, could transform this now-vast expanse of space into accessible areas for our species. And as we move into this new arena for expansion of earthkind, we must ensure that our own garbage will not be there to greet us.

12. Space Settlements

Clearly, in the past several years the possibility of large scale colonization of space has captured public imagination and interest. Interestingly, this surge of public excitement in space habitation was not sparked by NASA. Rather, through the work of Princeton University professor of physics Gerard K. O'Neill, a new option for our species has been suggested.

In his 1974 article in *Physics Today* ,(45) O'Neill brought both scientific and public attention to the possibility of creating large habitats in space within the next two decades. His initial studies suggested that space colonization would alleviate earth population problems, raise the quality of living standard of our species, create an industrial capability free of environmental damage, and encourage cultural diversity. O'Neill also believes that the colonization effort could eradicate warfare by lessening territoriality.

The size of the undertaking is bold by past measures of our space capabilities. Development of the space community, states O'Neill, could be realized using existing technology. Material for the first colony would come primarily from the moon, catapulted off the lunar surface by a "mass driver" to the colony site. Location of the community would be at the Lagrange libration points, L4 or L5, areas equidistant from the earth and moon. The initial colony is envisioned as housing some 10,000 people with a total mass of 500,000 tons.

Eventually, later colonies would be self-replicating and self-sufficient, constructed from quantities of both moon and asteroidal material. The colonization effort, believes O'Neill, could lead to space colony populations of 20 million people. Several thousand colonies could reside in the stable libration orbits. If such is the case, in less than 200 years more people might be living in space than on earth.

In its spin-produced gravity, the habitat would include controlled temperature and weather, using solar power as the colony's energy source. The largest colonies would involve a structure four miles in diameter and about twenty miles long. Such a colony would have a total land area of some seven thousand square miles.(46)

The interior of the colonies offers attractive settings, much like those of certain earth locations -- small villages, surrounded by forests, rivers, lakes, animals and birds. The colony would include special agricultural areas to feed the inhabitants. Bicycles and low-speed electric transportation could be used to cover the 20-mile distance of the cylinder interior. O'Neill suggests that recreational life aboard the habitat would include skiing, mountain climbing, sailing and other normal earth-like sports. Newly developed space activities would involve effortless man-powered flight in varying amounts of gravity and zero-gravity.

13. Reaction to Space Settlements

Since O'Neill's 1974 article, numerous technical evaluations have taken place.(47) A NASA design study of space settlements leads to the final conclusion that

> there seems to be no insurmountable problems to prevent humans from living in space. However, there are problems, both many and large, but they can be solved with technology available now or through future technical advances. The people of Earth have both the knowledge and resources to colonize space.(48)

Refinements of the O'Neill project have taken place over the years, particularly as to the function of future space colonists. What do 10,000 people do in space? Although O'Neill did initially refer to some space manufacturing, important for self-replication of the colony, it was clear that this would not justify large earth expenditures, perhaps upward to $100 billion for the first colony.

O'Neill was quick to learn that if the concept was to survive, colonization must include the production of some viable goods and services, beneficial to earth. By 1975, O'Neill suggested that space colonies could construct satellite solar power stations, to supply solar energy to earth via microwave.(49) These stations, built by the colony using non-terrestrial resources, would prove cheaper to construct at the colony site, compared to hauling material up from earth.

In 1976, O'Neill saw the colonization effort as "Pittsburgh at L2," and called for the creation of a space industrial center, using lunar material, and "unintentionally" creating a self-replicating space colony.(50)

Today, the space colonization effort is slightly subdued, brought about by what O'Neill calls "governmental myopia." His current research is centered on large scale manufacturing in space with no requirement for space colonies.(51) O'Neill has established a small, independent institute to support continued work.(52)

The space colonization theme has brought both criticism and praise. Some liken space colonies to the next evolutionary step for the species while others caution that colonization is a "betrayal" of human values.(53) A few comments contain the belief that space settlements involve problems which should be resolved first on earth.

Such is the case of Paul Csnoka, Director of the Institute of Theoretical Science, University of Oregon. Writing in *Futurist* , Csnoka warns that space colonization may lead to widespread oppression and violent global disaster. Pointing to numerous social and political problems based on historical analogy, Csnoka believes space settlement development should be limited until humanity becomes less violent. Restrictions on space colonization would be dependent on formation of a world government, capable of policing space. It is Csnoka's belief that large-scale colonization, without such a government, would be a form of suicide.

Concluding, Csnoka suggests the need to solve three problems prior to full-scale colonization: (1) how to settle conflicts (e.g., concerning the distribution of resources) nonviolently *and* justly, (2) how to safeguard the right of self-determination of various groups (on earth and in space colonies) without opening the door to perpetual turmoil, and (3) how to limit population growth and waste production to avoid finding ourselves in desperate situations leading to desperate actions.(54)

14. Artificial Worlds for Artificial People?

With space exploration and humanization of space gaining public acceptance, a most peculiar dichotomy may be developing. Is the production of "artificial worlds" counter to a growing social movement calling for a "back to nature," "small is beautiful," "soft-technology," and "decentralized" approach to problem solving? Will the merits of future space programs be measured in terms of "appropriate technology?"

It has been commented that the artificial living styles offered by space settlements are nothing more than the environment already produced by the American shopping mall. Such malls contain artificially independent environments, and are totally enclosed, self-contained, planned and controlled.(55)

Are space settlements comparable to putting a dimmer switch on the sun and stars? Perhaps a polarization of attitudes will take place with anti-colonization movement slogans of "artificial worlds for artificial people"?

The popularity of space settlements may be merely a comment on our generation -- besieged by a multitude of problems -- willing to trade in earth for space-based lifestyles. Or, on the other hand, perhaps colonization is but a necessary step that any technologically advancing civilization must take, assuring preservation of its species.

It is conceivable that space settlements may turn out to be merely space ships, capable of transporting large quantities of earthkind and equipment to other planets. Indeed, there exists the possibility that planetary engineering would permit a proliferation of our species, first to parts of our solar system, then outward to other star/planetary systems.

An early target for such a scheme would be Mars. Techniques for terraforming Mars were studied by NASA in 1975, concluding that "no fundamental, insuperable limitation to the ability of Mars to support terrestrial life has been unequivocally identified."(56) The report outlines possible mechanisms that could be used to transform Mars into a livable abode for earthkind. Among these are vaporizing the martian polar caps, thereby changing the martian surface temperature. By genetic engineering, oxygen-producing photosynthetic organisms could survive and grow on Mars, in turn creating a habitable atmosphere for our species. Mars may be only one possible target in our solar system for such engineering; perhaps Venus is a likely candidate as well.(57)

Whatever the outcome of space settlements, they offer us the luxury of knowing we can extend, in bodily form, the creative talents of numbers of our species into space. Though the idea of such colonization has captivated writers for many, many years, we are now capable of evaluating the concept on technical merits. But such a technological effort can only come to fruition if it reflects combinations of economic, political, and societal needs. To some, it has caused an early surge of future shock while there are those who see colonization as "the only road to save civilization.(58) It is clear that the possibility of space settlements has permitted us, perhaps prematurely, a peek into our future.

15. Conclusion

What lies ahead for earthkind is the growth of a social and legal order for the cooperative utilization of space. Through space involvement, we can become selective in creating and implementing new philosophies and institutions, permitting the growth of our species into space. Drawing upon our past history, this order must represent a fair and unrestricted flow of benefits to all on the global village earth, while lessening fears of exploitation.

Space exploration does not reside solely in the domain of the engineer. To insure an orderly and productive space program, multidisciplined decision-making will be required. To this end, innovative and creative educational mechanisms are required, extending traditional disciplines in the sciences and humanities to include the option of space.

But before earthkind can consider itself a member of what may be called a "galactic club of intelligent species," we must deserve membership. Today we find ourselves stumbling into the universe; inadequate legal regimes permit extensions of environmental carelessness, territoriality, and military ideology. In our rush to capitalize on the "zero-gravity dollar," are we failing to tap the ultimate resource of imagination, curiosity, and *exploration* ? Can a dollar value be placed on a search for extraterrestrial intelligence, or locating and understanding those cosmic oddities: black holes, quasars and pulsars?

There are those who argue that industries in space and off-earth drilling of asteroids and the moon will provide the base for an increased and outward program of exploration. This may prove true. Yet one can look at vast regions of the earth, primarily the oceans, where an explorer's eye has yet to visit, intent on *scientific* inquiry.

We have developed into a species capable of sailing to the stars, only to find ourselves breathing our last industrial revolution and shackled by governmental bickering and minimal programs of cooperation. Space could prove to be an "equalizer" for our species or a new divisive element of societal growth.

We can choose to escape the limits to growth or carry their essence into the universe. With our first 20 years of space exploration behind us, we are merely anchored in the harbor of near-earth, ready to set sail for the longer voyages ahead.

REFERENCES

1. For an overview of activities regarding the U.N. and outer space, see *The United Nations and Outer Space*, U.N. Office of Public Information, Sales No. E.77.1.9, New York, 1977.

2. David, Leonard, "The Military Uses of Outer Space,"presented at Twenty-Third Annual Meeting, AAS, October 18-20, 1977, San Francisco, California.

3. Robinson, George S., "Militarization and the Outer Space Treaty -- Time for a Restatement of Space Law." *Astronautics and Aeronautics*, February, 1978, pp. 26-29.

4. Kennedy, John F., *Public Papers of the Presidents of the United States*, January 1 to December 31, 1962, p. 669.

5. *World-Wide Space Activites*, Report prepared for the Subcommittee on Space Science and Applications of the Committee on Science and Technology, U.S. House of Representatives, Ninety-Fifth Congress, First Session, Serial G, September 1977. This document contains excellent material documenting the extent and range of international space planning.

6. Glaser, P.E., "Power from the Sun: Its Future." *Science*, Vol. 162, November, 1968, pp. 857-886.

7. Remarks by Brigadier General Stelling, Jr., Director of Space DCS/R & D, Hq. USAF, at the Twenty-First Annual Meeting of the AAS, August 26, 1975, Denver, Colorado.

8. Smart, General Jacob E., "Strategic Implications of Space Activities," *Strategic Review*, Fall 1974, p. 24.

9. Hansen, Lt. Col. Richard E., "Freedom of Passage on the High Seas of Space," *Astronautics and Aeronautics*, February 1978, pp. 76-83.

10. Haley, Andrew G., *Space Law and Government*, Meredith Publishing Company, New York, 1963, p. 157.

11. For a discussion of these terms and East/West definitions, see McNaughton, John I., "Space Technology and Arms Control," in *Law and Politics in Space*, Maxwell Cohen, ed., Leicester, 1964. Also Haley, Andrew G., *Space Law and Government*, Meredith Publishing Company, New York, 1963, pp. 154-155.

12. "Technology Seen Hindering Efforts to Limit Arms Race," *Washington Post* , November 29, 1976.

13. Wilson, George C., "Air Force Begins Development of Satellite Killer," *Washington Post* , September 23, 1977.

14. Covault, Craig, "Soviets Build Reusable Shuttle," *Aviation Week & Space Technology* , March 20, 1978, pp. 14-15.

15. Morgan, Lt. General Thomas W., "Space in the Air Force Future," taken from remarks to the American Defense Preparedness Association, October 15, 1975.

16. Sanborn, Colonel Morgan W., "National Military Space Doctrine," *Air University Review* January-February, 1977, pp. 74-79.

17. "Satellite Killers," *Aviation Week & Space Technology* , June 21, 1976, p. 13.

18. "Soviets See Shuttle as Killer Satellite," *Aviation Week & Space Technology* , April 17, 1978, p. 17.

19. Brodie, Bernard and Fawn M., *From Crossbow to H-Bomb* , Indiana University Press (Canada), 1962, p. 178.

20. Douglas, John H., "High Energy Laser Weapons," *Science News* , July 3, 1976, pp. 11-13.

21. *Summer Workshop on Near-Earth Resources* , James Arnold and Michael Duke, eds., NASA Conference Publication 2031, Conference date August 6-13, 1977; published 1978.

22. O'Leary, Brian, "Mining the Moon and Asteroids and Living in Space," *Astronautics and Aeronautics* , March 1978, pp. 21-23.

23. Glaser, Peter E., "Solar Power Satellite Developments," presented at 1978 Goddard Memorial Symposium, March 8-10, 1978, Paper No. AAS 78-022.

24. Bekey, I., and Mayer, H., "1980-2000: Raising Our Sights for Advanced Space Systems," *Astronautics and Aeronautics*, July/ August 1976.

25. *Seafloor Engineering: National Needs and Research Requirements,* Marine Board, National Academy of Sciences, Washington, D.C., 1976.

26. For an overview of the issues involved in seabed mining, see *Ocean Manganese Nodules* (second edition), prepared by the Congressional Research Service for the Committee on Interior and Insular Affairs, U.S. Senate, Washington, D.C., February, 1976.

27. Arnold and Duke, *op. cit.*

28. *The United Nations and Outer Space op. cit.* p. 9.

29. Smith, Philip, "Prospects for International Cooperation on the Moon: The Antarctic Analogy," in *Man on the Moon -- The Impact on Science, Technology, and International Cooperation* , Eugene Rabinowitch and Richards Lewis (eds.), Harper and Row, New York, 1969, pp. 85-98.

30. Smith, Philip and Johnson, Rodney, "Antarctic Research and Lunar Exploration," in *Advances in Space Science and Technology* , Vol. 10, Academic Press, New York, 1970, pp. 1-44.

31. Nossiter, Bernard D., "Antarctic Oil -- Who Can Exploit It?" *Washington Post* , September 20, 1977.

32. Gehrig, J.J., "Geostationary Orbit -- Technology and Law," presented at XXVII International Astronautical Congress, Anaheim, California, October 10-16, 1976. Paper No. IAF-ISL-76-30.

33. Hupe, Howard, "Carving Up the Geosynchronous Orbit," *Astronautics and Aeronautics* , February, 1978, pp. 10-15.

34. Gorove, Stephen, "Solar Energy and Space Law," *Studies in Space Law: Its Challenges and Prospects* A.W. Sijthoff-Leyden, 1977, pp. 205-209.

35. "DOD Guidelines for Technology Export," in *World-Wide Space Activities* , *op. cit.,* p. 25.

36. Cole, Dandridge M., "Response to the 'Panama Hypothesis'," *Astronautics,* June, 1961, pp. 36-39.

37. Kessler, Donald J. and Cour-Palais, Burton G., "Collision Frequency of Artificial Satellites: The Creation of a Debris Belt," *Journal of Geophysical Research.* June 1,1978, pp. 2637-2646.

38. Simpson, J.P. and Witteborn, F.C., "Effect of the Shuttle Contaminant Environment on a Sensitive Infrared Telescope," *Applied Optics* , August 1977, pp. 2051-2072.

39. Stein, Harry G., "The Third Industrial Revolution: The Exploitation of the Space Environment," *Spaceflight* , September 1974, p. 330.

40. O'Leary, Brian, "Mining the Apollo and Amor Asteroids," *Science* , July 22, 1977, p. 365.

41. O'Neill, Gerard K., "The Low (Profile) Road to Space Manufacturing," *Astronautics and Aeronautics* , March 1978, pp. 24-32.

42. Davis, H.P., "Technology Challenges in Deploying and Using Mass Driver Systems," *New Moons -- Towing Asteroids into Earth Orbits for Exploration and Exploitation* , Lunar Science Institute, Houston, Texas, March 16, 1977, pp. 6, 13.

43. Nossiter, *Op. cit.*

44. Russell, Mary, "House Votes Alaska Wilderness Bill,"*The Washington Post* , May 20, 1978.

45. O'Neill, Gerard K., "The Colonization of Space," *Physics Today* , September, 1974, pp. 32-40.

46. O'Neill, Gerard K., *The High Frontier -- Human Colonies in Space* , William Morrow and Company, Inc., New York, 1977.

47. *Space Manufacturing Facilities -- (Space Colonies)* , Proceedings of the Princeton/AIAA/NASA Conference, Jerry Grey (editor), American Institute of Aeronautics and Astronautics, Inc., New York, March 1, 1977.

48. *Space Settlements -- A Design Study* , Richard Johnson and Charles Holbrow (editors), NASA, SP-413, Washington, D.C., 1977.

49. O'Neill, Gerard K., "Space Colonies and Energy Supply to Earth," *Science* , December 5, 1975, pp. 943-947.

50. O'Neill, Gerard K., "Engineering a Space Manufacturing Center," *Astronautics and Aeronautics* , October 1976, pp. 20-36.

51. O'Neill, Gerard K., "High Frontier -- Technical Progress, A Resolution, Commitments," *Astronautics and Aeronautics* , March 1978, pp. 18-20.

52. Space Studies Institute, Box 82, Princeton, New Jersey 08540.

53. For various viewpoints on space colonization, see *Space Colonies* , Steward Brand (editor), Penguin Books, New York, 1977.

54. Csnoka, Paul L., "Space Colonization: An Invitation to Disaster?", *The Futurist* , October 1977, pp. 285-290.

55. Kowinski, William S., "The Malling of America," *New Times* , May 1, 1978, p. 52.

56. *On the Habitability of Mars -- An Approach to Planetary Ecosynthesis* , M.M. Averner and R.D. MacElroy (editors), NASA, SP-414, Washington, D.C. , 1976.

57. Oberg, James, "Terraforming,"*Astronomy* , May 1978, pp. 6-25.

58. Asimov, Isaac, "Here's the 'Only Road' to Save Civilization," *Science Digest* , February 1977, pp. 8-12.

4

Models of
Long Range Growth

William A. Gale and Gregg Edwards

Abstract

This paper reviews the possibilities of colonizing space. The physical resources of the solar system might support a sextillion humans. This development would require five millennia at 1/2% per year growth rate. Interstellar colonization may spread through the galaxy at about a tenth the speed of light. The velocity of intergalactic expansion may be between that and the speed of light.

A model with multiple centers of growth is presented. The probability that growth originates from a given star is related to the prior probability that extraterrestrial life would have colonized the solar system by now. Limits are established on the probability of interstellar growth originating at a given star. The upper limit is one per thousand galaxies. The expected number of settlements visible, their distances, and their diameters are calculated as functions of intergalactic expansion velocity and prior probability of the solar system having been colonized by now. If the expansion velocity is less than half light speed, other settlements could well be visible. For expansion near light speed, no other settlements will be visible.

Models with fundamentally different assumptions are the communicative models, and the low impact expansive models. The models presented in this paper differ from these models in allocating intelligence a major role in the future development of the universe, and in offering a much greater prospect for future growth for our descendants.

72 *Gale and Edwards*

1. Introduction

The limits of terrestrial surface growth are being reached. But for Technological life, resources beyond the terrestrial surface may be available. These include sunlight, of which the earth intercepts the half of a billionth part, and materials of which the earth has the thousandth part in the planetary system. Where are the limits to growth if these resources can be developed?

The paper presents models in which resources are completely developed for direction by intelligence. We note that life on earth has developed niches on most of the earth's surface, expanding as it was able to. For instance, with the development of multicell beings, the land areas could be colonized, and thereafter land-dwelling beings are found.

Thus, the first assumption of the models presented is:

1.   Technological life will expand to use all the resources available to it.

For expository purposes, we distinguish two different levels of feasibility of expansion. We characterize an expansion as "feasible" if its initiation requires 10^{-3} of the total currently available resources, and as "unavoidable" if its initiation requires 10^{-9} of total currently available resources. The feasible level is based on observing that the Apollo project took that fraction of American resources during its lifespan. The estimation of unavoidability is probably conservative, because action by dissidents has frequently involved much larger portions of available resources. The Mayflower voyage, for instance, carried 10^{-5} of Elizabethan England's population.

We will also discuss (and the models will assume) the propositions:

2.   Complete development of a solar system's resources, including use of most of the star's radiated energy, is possible; and

3.   One way to travel to stars and galaxies is possible.

When these assumptions hold, volume inhabited by intelligence is markedly different from a "wild" volume. For instance, its radiation will include very little visible light, but will be mainly in longer wavelengths.

Hence a question to be answered by these models is why the universe appears wild (explainable without recourse to intelligence). We sketch the kinds of phenomena that are expected to be observed. It is possible, within the uncertainty of the parameters of the models, that evidence of settled volumes can be found. We return to this question in the discussion at the end.

For comparison with other work, it is important to distinguish communicative technological life and expansive technological life. Communicative technological life has the capability to communicate across interstellar distances. Expansive technological life has the capability physically to expand across interstellar distances. Previous estimates of the frequency of technological life were directed to communicative technological life. This paper develops a model which places upper and lower limits to the probability (per star) for the emergence of expansive technological life. The establishment of limits allows us to define "high estimates" for the probability of technological life as those for which the estimate exceeds our upper limit. The previous estimates for communicative technological life have satisfied:

4.   The probability of technological life originating is greater than one per thousand galaxies.

An important result of this paper is that at least one of the above four assertions is false. Yet all are appealing. The paper provides a framework for their joint consideration. The importance of the framework is that it highlights not just what must be asserted by different positions, but also what must be denied. Progress may be made by careful consideration of the arguments against each proposition. Classes of models having distinguishable observational consequences follow from the rejection of different assertions.

The paper first discusses the first three assumptions in some detail as a single center growth model. This discussion identifies two parameters as most uncertain. But in an open universe, what can happen at one point is certain to be repeated at some distance elsewhere. Thus growth initiated at other points limits the growth of any one settlement. Therefore the paper develops a multicenter growth model, based on the two parameters identified in the single center growth model. We review other models which take different approaches to the assertions above. A final section compares the predictions of the different models, and discusses their implications.

2. Single Center Growth Model

The multicenter growth model assumes that each center follows the same development. This section discusses the possible growth from one center. The discussion identifies two basic parameters which are poorly known at best: (1) the probability per star of originating interstellar growth, and (2) the expansion velocity possible for interstellar or intergalactic growth.

In the one case of life origination that we are familiar with, the process was apparently local spontaneous generation. The directed panspermia hypothesis (Crick and Orgel, 1973) poses an alternative. The suggestion was that the anomalous abundance of molybdenum in organisms might indicate origination elsewhere than earth. The transport mechanism would be the intentional dispersion of specially formed and protected cells. Gualtieri (1977) showed that considering all trace elements, the concentrations in bacteria, fungi, plants, and land animals are strongly correlated with concentrations in sea water. Deviations arose primarily from the chemical natures of the elements. We suppose that life and technological life arise spontaneously about some stars.

The probability of the spontaneous generation of life is not known. It is ultimately the probability that sufficient conditions for life to form and develop are present at the times required. The sufficient conditions are not known, however. Some of the prerequisites have been identified, such as solar mass limits (Huang, 1960), and size of secondary object to the star (Hohlfield and Terzian, 1977). Hart (1978) has pointed out the stringent temperature requirements. Numerous experiments starting with Miller (1953) have shown that organic precursors, such as amino acids, are easily formed from the inorganic constituents expected on the primitive earth. There are, however, only schematic theories of the process by which the genetic code was formed (Crick, 1968; Orgel, 1968). The temperatures (Miller and Orgel, 1974), substrates (Anderson and Banin, 1974; Ostroshchenko and Vasilyeva, 1977), and energy sources (Bar-Nun et al., 1970; Sagan and Khare, 1971) available and required are still under active debate. Crick (1973, p.52) has stated, "It is not possible at the moment, with our knowledge of biochemistry to make any reasonable estimate whatsoever of the factor f_1 [probability of life forming]."

Therefore, the probability of expansive technological life arising around a given star is unknown. For the generation of life and evolution to technological competence is one way to generate technological life. Not knowing the probability of a key step for this particular way, we could not give the overall probability even if we knew the probability of all other ways of forming technological life. Thus we denote this probability by p, and regard it as a parameter. It can have any value from 0 to 1.

We do have one estimate of the time for technological life to emerge after star formation, namely the age of the earth. This has been estimated to be 4.5GY (York and Farquhar, 1972). This is presumably one number from a distribution. However, with one drawing only, we have an estimate of the mean of the distribution, and none of the standard deviation. It should be noted that while we may be described as a communicative technological life, we are not yet an expansive technological life. Thus we do not have even one observation on the relevant population. However, we argue below that the required technology, if feasible at all, will be developed within the next few thousand years. This is a negligible addition to 4.5GY.

The assumption that life will expand if it can is conservative in the sense that this predicts continuity rather than change of behavior. It is also conservative in that the greater likelihood may be overexpansion, especially in the first areas developed. We note that the centers of development of civilization in the Middle East were long ago overpopulated. This may be the price of learning how to develop a given resource size.

The first technological extension required is the ability to completely use the resources of a solar system. We currently use only the resources, including the biosphere resources, of the surface of one planet. We do seem to be approaching the limits of these resources (Meadows et al., 1972) though how closely is debated (Kahn, 1976). The technology required to use other resources in the solar system is suggested by the concept of space colonies (O'Neill, 1977; Heppenheimer, 1977). The basic concept is to lower the required mass per being by living inside of many small mass objects rather than on the outside of a few large mass objects. Their continued use requires being able to construct an ecologically closed colony or network of colonies, stable over periods of time compatible with the lifetime of the sun, i.e., gigayears. They might need to be designed assuming the possibility of violent conflict, because the alternatives available in differing internal colony social systems and in differing levels of cooperation between colonies offer fertile grounds for conflicts. The proliferation of space colonies could lead to a swarm completely surrounding the sun. We call this a swarm sphere.

Dyson (1966) made several points relevant to the feasibility of a swarm sphere. He pointed out that rigid structures of up to a million kilometers diameter could be built in space, and that they would remain light enough to build the required number. He also pointed out the feasibility of taking Jupiter apart. Extensions of the mass-driver suggested by O'Neill (1977) for disassembly of airless bodies may make a major contribution to planetary disassembly. The swarm sphere is one possible means of completely developing the solar system. More advanced technology might allow the construction of a "band sphere." A series of bands would cover the surface of a sphere more efficiently than random particles. The technology to hold a spinning band together is not available now, however.

The carrying capacity of the sun is about 10^{21} (a sextillion) human beings. The mass of the planetary system would allow 2×10^6 kg for each human, including 200 kg of oxygen, and 10^4 kg iron. We express growth here and elsewhere in terms of human beings for the sake of exposition. Although there is no certainty that the beings would be human or even protoplasmic, there is also no necessity of a form change in order to continue growth.

Within a few thousand years, the solar system could be completely developed both in matter controlled and bits of information stored. The worldwide average growth rate for number of humans in the half-millennium since the European colonization of the Americas began has been about 1/2% per year (Pop. Ref. Bur., 1976). This growth rate allowed room for major depressions, pestilences, and wars. If this average held until 10^{21} humans existed in the solar system, the time required would be about 5400 years. The stages of this development might be labeled approximately by the resources being exploited as (1) asteroids, 1000 years; (2) moons of major planets, 1200 years; (3) terrestrial planets (excepting earth and moon), 2200 years; and (4) major planets, 1000 years. These estimates are based on the assumed growth rate and tabulated masses of the solar system bodies. The major planets would sustain growth only as long as the asteroids would because of the greatly increased population base.

The limits of information storage are more crudely set. Assume that 10^6 nucleons will be required to store one bit, and that 10% of nucleons can be devoted to information storage. We estimate that currently the greatest information storage is in the human brain, estimated at 10^{12} bits per brain (Minsky, 1973), or 4×10^{21} bits. Then stored information could grow exponentially at 1% per year for 6000 years. An error of three orders of magnitude in growth possible or a ten per cent error in growth rate corresponds to a 10% error in years to achieve the growth. After the bit limit is reached, increases in knowledge will require improvement of coding efficiency. As an approximation we will suppose that knowledge growth will be proportional to physical growth after this point.

Thus the 10^{-9} part of a completely developed solar system might be 10^{12} humans. They would live in ecologically closed systems needing only propulsion to be adequate vessels for stellar travel. The greatest part of their resources would be hydrogen which we assume would be usable as fuel following knowledge development of a few thousand years. Thus the project of going to another star would be possible for any 10^{-9} part of the total swarm sphere. Under the assumption that the lower rotation rates for stars F0 and later indicates a secondary system (whether planetary or not), most stars will have resources to support exponential growth for 10^{12} humans for many orders of magnitude. Thus the project of interstellar travel is a growth project that can be undertaken by the 10^{-9} part of a swarm sphere. Accordingly, we deduce that if the solar system is completely developed then interstellar travel is unavoidable. The development of the solar system is only feasible, however. It may be decided not to undertake the development.

The velocity with which colonization takes place depends on the travel velocity, v, and on the regeneration time, t (time to build a new fleet and the population to travel in it). Suppose settlement points average a distance d apart, and define travel time as d/v. If the nearest points are settled, the average expansion velocity is

$$v_e = \frac{d}{t + d/v}. \tag{1}$$

Unless regeneration time, t, is less than travel time, d/v, v_e will be reduced by at least a factor of 2 from v. However, if t is comparable to d/v, its effects can be greatly reduced by settling points an average Nd apart, and then filling in the intermediate spaces in parallel. The expansion velocity becomes

$$v_e = \frac{Nd}{t + Nd/v} = \frac{d}{t/N + d/v} \tag{2}$$

If t is comparable to d/v, then t/N can be made small compared to d/v

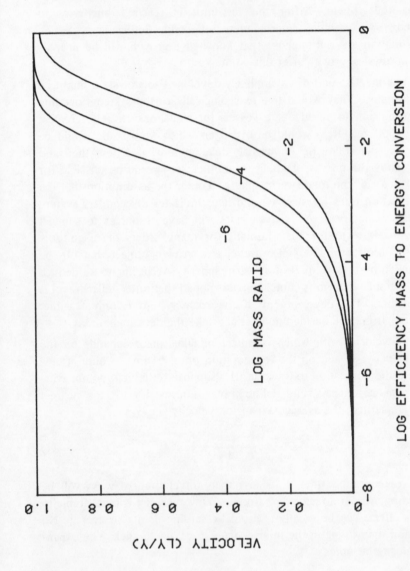

Figure 1   Rocket Equations

for a modest sized N. Another strategy would start with a fleet large enough to settle N points, and have portions of it stop along a ray. In examining how much regeneration might slow expansion velocity, Jones (1976) found that for reasonable ranges of parameters, expansion velocity was not less than one-tenth the travel velocity. Here we are asking how rapid the expansion might be, for to the swift will accrue greater resources and they will thus dominate the interstellar ecology.

Travel velocity can be estimated from the relativistic rocket equation

$$v = \frac{1 - (M_v/M_o)^{2w}}{1 + (M_v/M_o)^{2w}} \tag{3}$$

$$w = (2e - e^2)^{1/2} \tag{4}$$

where

v = velocity achieved, as a fraction of lightspeed

M_v/M_o = ratio of mass at velocity v to initial mass

w = exhaust velocity as a fraction of lightspeed

e = fractional efficiency converting fuel mass into energy

This equation describes the result of acceleration. It remains to decelerate at the destination. To deliver the final mass M at rest at the destination, from the initial mass M_o at rest at the origin, we use $(M_v/M_o) = (M/M_o)^{1/2}$ for both acceleration and deceleration, giving

$$v = \frac{1 - (M/M_o)^w}{1 + (M/M_o)^w} \tag{5}$$

We term the ratio M/M_o the delivered mass ratio. This equation is illustrated in Fig. 1 and in the following table, which shows the velocities that result for several combinations of M/M_o and two different estimates of a feasible w. The lower w shown, .03, is Heppenheimer's (1977) estimate for a foreseeable rocket technology. The higher w shown results from assuming a 50% efficient fusion-based drive. (See Table 1).

Table 1
Fusion Rocket Velocities

delivered mass ratio	exhaust velocity	
	w = .03	w = .084
10^{-1}	.035	.10
10^{-2}	.069	.19
10^{-4}	.14	.37
10^{-6}	.20	.52*

Heppenheimer (1977) concluded that with w=.03 velocities about .1 were reasonable for interstellar travel. The delivered mass ratio 10^{-3} may be reasonable for interstellar travel. One might expect improvement to a few tenths of light speed for the delivered mass ratio 10^{-3}. Figure 1 plots the velocity v attainable as a function of $\log_{10}(M/M_o)$ and of $\log_{10}e$. The values of w=.03 and .084 correspond to e=.00045 and .0035. Completely efficient fusion of H to He has e=.007. The figure shows that if e=0.1 can be attained by some technology (far beyond our horizon), then very high relativistic velocities are possible.

The second determinant of expansion velocity is regeneration time. This is quite uncertain, and would be long with respect to travel time if it relied on dissidents at each stage. Once started, it is perhaps more reasonable to suppose that an expansion philosophy would dominate and regeneration would be done as quickly as possible. This might be after a factor of ten growth from the settling fleet. Because it would not be the first time for such growth, a rate larger than 1% per year might hold, perhaps 3% per year. Then a factor of ten growth would take 60 years. This is on the order of travel time between nearest stars at v = .1, so for interstellar expansion the expansion velocity can be kept up to the travel velocity as discussed above. Thus the galaxy might be settled in about the time it would take to cross it. At v = .1, this would be about one million years. This result agrees with Hart's (1975) statement. One of the results in the next section quantifies his conclusion that it is therefore unlikely that other (expansive) technological life

---

*This figure can be extended to v=0.60 if original acceleration is provided for by a fixed linear accelerator.

exists in this galaxy. Because the density of stars increases toward the center of the galaxy, a plot of number of stars against distance from earth shows a cubic dependence despite the flatness of the galaxy. Thus growth through the galaxy will be approximately cubic in time.

As the galaxy is settled, we might expect considerable rearrangement of the matter of the galaxy. As Dyson (1966) observed, it may be judged that too much matter is tied up in stars. He pointed out the possibility of extracting matter from stars by arranging collisions. Another way might be to reflect the light of the star back into it. Louvered mirrors mounted on an encircling swarm could achieve a variety of energy modulations in a star.

The galactic matter could be arranged into a single structure a few light years in diameter to facilitate communication. (The size is set by the surface gravity and the strength of materials. A few light year diameter results for 1 g surface gravity.) The time scale would be a hundred to a thousand times that of the initial settlement because only a small fraction of the total mass could be used for acceleration and deceleration of the material. If the project would be done with only one acceleration and deceleration for each mass involved, then expenditure of 1% of the mass would allow the construction to take place in about one GY. Because the human form might be abandoned, we cannot specify the temperature that would be chosen. Lower temperatures than we live at would allow greater use of energy. Likewise, the intensity of energy use might be less than its rate of dissipation by wild mechanisms. For, by using less energy per unit time the lifetime available would be extended.

The resources available for intergalactic travel would be vast. When 1% of the galaxy had been settled, the 10^{-9} part of total resources would be one solar mass. This point would be reached in 170,000 years if cubic growth would settle the galaxy in 1 million years. The resources of one solar mass would allow a delivered mass ratio of 10^{-6} for a delivered mass equal to the earth's mass. This delivered mass could be configured to support 10^{18} people. Therefore these models assume that intergalactic travel is feasible.

The velocity of intergalactic expansion should be at least v/c=.1, and possibly v/c near 1. If a delivered mass ratio of 10^{-6} can be used with 50% efficient fusion, then v/c=0.5 would be attainable. If a fuel with e=1 (storable antimatter) could be produced with an efficiency of 5% from fusion energy,then using one solar resource to produce fuel

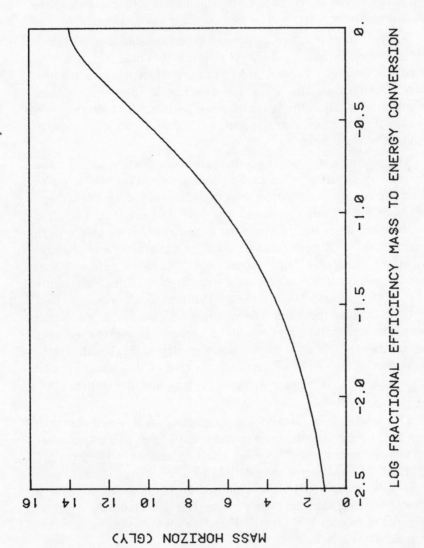

Figure 2   Mass Horizon Equation

would allow travel at velocity $v/c=0.997$ for a delivered mass equal to earth's. Considering the knowledge to be had in a society commanding 10^{56} bits of information, we do not feel that intergalactic travel velocities close to light speed can be excluded. The landing pattern could place 10^{15} people at each of 10^3 stars. A few of them could be designated as regeneration points. The regeneration time would be that to organize a billionfold of material, which at 3% growth would require 600 years. The regeneration time would thus be negligible compared to even the few million (or tens of millions at $v/c=0.1$) years travel time for travel between closely spaced galaxies such as our own and M31. It is even smaller compared to time required to travel the more typical distance of 30 MLY. Hence the expansion velocity would almost equal the travel velocity.

The limit to the matter that can be collected in one place is set by the expansion of the universe. We call the mass horizon the distance at which half the mass can be stopped from expanding away from some center. The non-relativistic calculation of the mass horizon is as follows. The velocity to be overcome is proportional to distance:

$$v = .05 \, r$$

where r is the distance in GLY. To stop this with rockets having $M_v/M_o = 0.5$, we have also

$$v = \frac{1 - (\frac{1}{2})^{2w}}{1 + (\frac{1}{2})^{2w}} \approx w \ln 2 \qquad (6)$$

Equating the two expressions gives

$$r_{1/2} = \frac{\ln 2}{.05} \, w = 14 \, (2e - e^2)^{1/2} \qquad (7)$$

where $r_{1/2}$ is the mass horizon in GLY, and e is the basic energy conversion efficiency for matter in the galaxy. The mass horizon depends only on this energy efficiency. Figure 2 shows $r_{1/2}$ as a function of $\log_{10}e$. For 90% efficient fusion ($e=.0063$), the mass horizon is 1.6 GLY. If there is antimatter in the universe, e near 1 is possible. A relativistic version is needed in this case. A black-hole rocket which might liberate a great energy fraction but would not allow all of the fuel mass to be ejected, would give results equivalent to a moderate e.

For a single center of growth in an open universe there is no limit to growth. A time would come when the new galaxies approached would be burned out stars of various kinds, but even these would provide resources for an advanced technology. The timespan of intelligent life is under its own control. By having fewer beings, or using less energy per being, the timespan is extended. However, in an open universe there is an infinite volume in which by chance other settlements may also originate. In an open universe the other settlements are thus infinite in number. The model of the next section addresses the likely limits to growth set by other settlements.

3. Multicenter Growth Model

The previous section sketched the possible growth from one center and identified as basic uncertain quantities the probability of spontaneous generation of a center, and the velocity of expansion. This section develops a model of expansion from many centers taking these two quantities as parameters. The model approximations are discussed, followed by the calculation of several quantities of interest.

We approximate the rate of generation of stars as uniform in space and time. The rate of star formation in our galaxy is believed to have declined perhaps by a factor of two during the last 10 GY. The spatial distribution is only roughly uniform within galaxies, very concentrated in galaxies compared to intergalactic spaces, and approximately uniform when distances between galaxies can be regarded as small. The rate of generation of stars is denoted by k, with units $(time)^{-1}$ and $(distance)^{-3}$.

We assume that stars began their spontaneous generation exactly T_G time ago. The galaxies are believed to have begun their generation about 10 GY ago (Sandage, 1972), which we take for T_G. This number may be uncertain by 30% due to uncertainties in estimating other cosmological parameters. Using this time, the estimate 5×10^{-32} gm/cm^3 (uncertain to an order of magnitude) for the density of stellar matter in the universe, and assigning 2×10^{33} gms as the mass of each star, we calculate $k = 2 \times 10^{15}$ star GY^{-1} GLY^{-3}. By a change in parameter discussed below, most of the results are insensitive to this numerical value.

We assume that the development time for expansive technological life is always the same as it has been for earth to develop a communicative technological life. That is, we are approximating the distribution of development times by our (poor) estimate of its mean. We do not have any estimate of the distribution's standard deviation. By treating the standard deviation parametrically, its impact could be established. Because the distribution conditional on development time less than T_G is often the actual distribution to use, and because the standard deviation of this conditional distribution cannot be large with respect to the estimated mean, we believe the effect of this approximation is not large. Errors in the estimation of the mean are partially removed by a change of parameters discussed below. This time is denoted T_S and is estimated to be 4.5 GY, being perhaps in error by 10% (York and Farquhar, 1972).

We assume that with probability p per star, any given star will have expansive technological life spontaneously generated around it. After the period T_S from the time of the star's formation, any technological life starts expanding at the uniform rate v. Thus at any moment after expansion starts, the volume occupied is spherical.

We now calculate the probability p^* of expansive technological life not having arrived at some point of space-time. Because the model is spatially uniform, we can select a coordinate system with origin at the given point. (Later we will identify the origin as our location.) Denote the time in question by t. If t is less than T_S, there is no expansive technological life in the universe, and so $p^*(t<T_S)=1$. If t is greater than T_S then we reason as follows. The probability of expansive technological life not having reached the origin by time t is the probability of its not having originated on any of the stars from which it could have settled the origin by time t. The probability of not originating at a given star is 1-p. For small p, this quantity is close to e^{-p}. If the number of stars from which the origin could have been settled by time t is $N(t)$, then the probability of expansive technological life not originating on any of them is the product of the probabilities of not having originated on each of them:

$$p^*(t) = (e^{-p})^{N(t)} = e^{-pN(t)}. \tag{8}$$

We have now to calculate the number of stars from which the origin could have been settled by time t. Expansive technological life originating at distances $r > (t-T_S)v$ could not have reached the origin by time t. For a star at distance r, the star could have been formed at any time before $t-T_S-r/v$ and expansive technological life forming on the star would reach the origin. In the spherical shell of thickness dr and radius r, there is volume $4\pi r^2 dr$, and hence

$$dN(t)=4\pi kr^2(t - T_S - r/v) \, dr \tag{9}$$

stars from which the origin could have been settled. Figure 3 plots dN/dr against r/v. The maximum density occurs many GY travel time away, the scale being set by the magnitude of $t-T_S$.

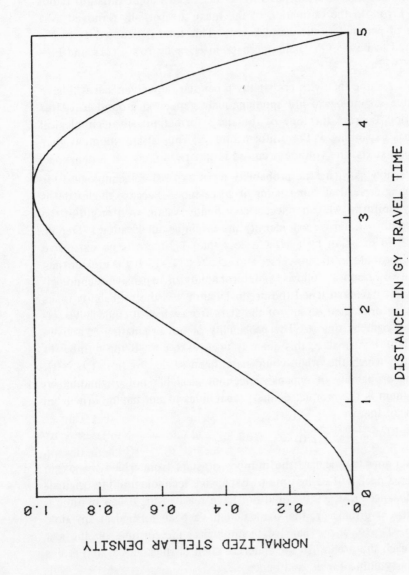

Figure 3 Density of Stars from Which Expansion Could Have Reached Here by Now

The total number of stars from which the origin could have been settled is found by adding together all the spherical shells out to radius $v(t-T_S)$:

$$N(t) = \int_0^{v(t-T_S)} 4\pi kr^2 (t - T_S - r/v) \, dr \tag{10}$$

$$= \frac{4\pi k}{12} v^3 (t-T_S)^4 \tag{11}$$

Notice that the result depends only on $(t-T_S)$. This is because the assumption of constant development time is equivalent to calculations assuming zero development time, but with T_S subtracted from each time coordinate. The number of stars calculated above is proportional to the space-time volume of a cone extending backwards in time from $t'=t-T_S$ to $t'=0$ and bounded by the surface $t'=T-r/v$. We call this the velocity v cone. Also we will denote $T=T_G-T_S$. Then the probability that expansive technological life would reach here by now is

$$1 - p^* = 1 - \exp(-pkv^3T^4) \tag{12}$$

where $4\pi/12$ has been taken as 1. Figure 4 plots $1-p^*$ as a function of v and $\log_{10}p$. Observe that for the velocities shown only $10^{-20}<p<10^{-14}$ gives $.01<p^*<.99$. Jones (1978) presented a similar calculation for expansion within a galaxy and concluded that if interstellar colonizing civilizations were common, their lifetimes must be very short. We have supposed long lifetimes and infer rareness.

The formula above views p and v as basic parameters, and allows the calculation of $p^*(t)$. However, the range that can be described for $p^* = p^*(T_G)$ is much more restricted than the range that can be described for p, so we will use the formula to eliminate p in favor of p^*. For one limit on p^*, we observe that colonization of the highly observable kind supposed in this model has not reached the solar system (nor even the galaxies in the local group). Hence it is very unlikely that the a priori probability of this galaxy being settled by now was as high as .99. Thus $p^*\geqslant.01$ is one reasonable limit. We would interpret $p^*=.01$ as a case of our being relatively late to develop as a center of settlement. Likewise, it would be unlikely for us to be very early in our emergence. Given that we have emerged, we suggest that $p^*\leqslant.99$ is another reasonable limit. We interpret $p^*=.99$ as a case of our being early to develop. Likewise, we describe $p^*=0.5$ as a case of middle emergence time. From this viewpoint we calculate p as a function of p^* and v:

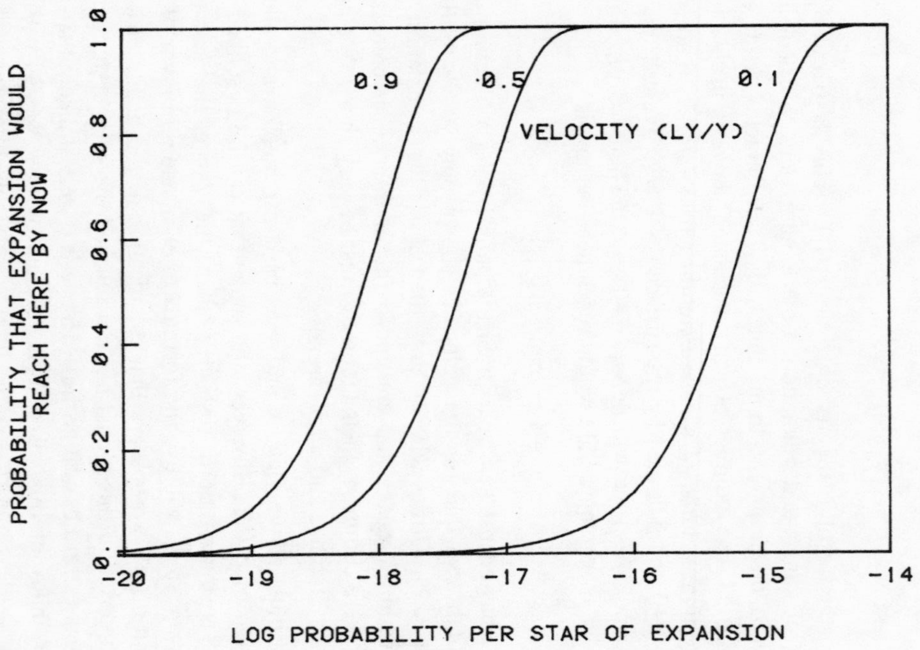

Figure 4 Limits to Probability of Expansion

$$p = (-\ln p^*) \frac{12}{4\pi} k^{-1} v^{-3} T^{-4}. \qquad (13)$$

The table at the end of this section gives numerical values. By asserting a range on p^* and v, we thus assert a range for p. The highest value of p is 4×10^{-4} per galaxy, the lowest is 1×10^{-9} per galaxy (where "galaxy" is a convenient unit, taken as 10^{11} stars). Notice that each of these numbers is as uncertain as the estimate of the density of stellar matter in the universe, which could be an order of magnitude. When we substitute the above expressions for p, dependence on k and this uncertainty are removed. Thus we view the model as setting a range for the probability of spontaneous generation of expansive technological life.

The predictive content of this model is that large scale settlements may be visible. However, their observation is not easy. The primary feature will be a lack of ordinary galaxies in an approximately spherical volume. The radiation from the interior of the sphere should be approximately black body in frequency distribution. However, its temperature might be quite low if a crystalline form of being were adopted. Also its intensity might be considerably less than the intensity of the replaced galaxies, the beings having chosen to be fewer in number for a larger period of time. Over longer periods of time, the "disappearance" (at visible frequencies) of ordinary galaxies would be a definite signal of settlement.

The absence of visible galaxies in a volume of space can be determined from a catalogue of visible galaxies showing three dimensional location. However, while the two angular dimensions of astronomic location can be easily determined with great accuracy, the radial distance is difficult to determine accurately, and is not yet available for a sufficiently large catalogues of galaxies. Also, surveys to the required distance of several GLY are available only for the northern hemisphere. One observation can currently be related to the model. If volumes of space with no visible galaxies were common, then this would imply an anti-clustering of galaxies (Davis, 1976). Since clustering, on average, is indicated by the observed correlation functions (Peebles and Groth, 1975; Bahcall, 1977) we need only consider the case of few or no visible settlements.

The following paragraphs present some model calculations of the number, size, distance, and expansion rates of possibly observable settlements. We calculate these observable quantities in the non-interacting approximation: each settlement remains spherical and

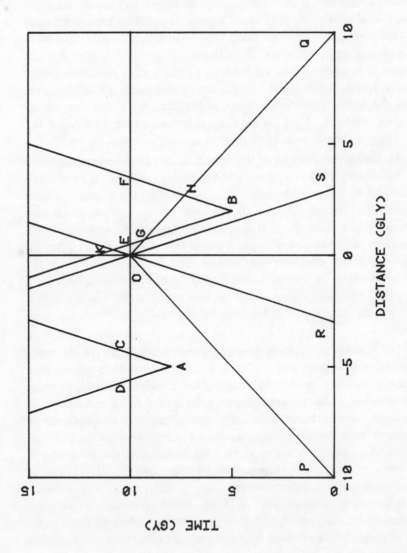

Figure 5   Diagram of Velocity Cones

preempts no others. The approximation leads to results whose error becomes large when p* becomes small (late formation); however, the absence of widespread settlements can be taken as known. Furthermore, the calculation is exact for the probability that at least one settlement is visible.

Figure 5 illustrates the calculation for how many settlements are visible from the origin, given that none have arrived by now. It shows the time dimension vertically and one space dimension horizontally. It is a section through the origin of a four dimensional space. The point marked 0 is here, now. A velocity v cone is shown as ROS. This is the cone from which settlements could have reached here by now and which is excluded in the current calculation. The light cone (the velocity 1 cone) is shown as POQ. Settlements originated within this cone would be visible now. The volume interior to POQ but exterior to ROS is to be calculated. Point B indicates a settlement which would be visible by now (G and H showing the visible edges toward and away from here). The settlement originated at B would not reach the origin until the time at point K. Point A shows another settlement which might exist now (extending from D to C) but as yet is not visible.

The settlements observable from the origin, given that no settlement contains the origin, are in the volume outside the cone bounded by t'=T-r/v and inside the cone bounded by t'=T-r (where we chose units for velocity such as GLY/GY, such that c=1). For settlements originating within the velocity v cone would contain the origin by time T, while the light signal from settlements originating outside the velocity 1 cone would not yet have reached the origin. The distribution of settlements, in the non-interacting approximations, is proportional to the number of stars between the cones and is described by

$$dN = \begin{cases} 4\pi kr^2((T-r)-(T-r/v)) \, dr & 0 < r < Tv \\ 4\pi kr^2((T-r)-0) \, dr & Tv < r < T. \end{cases} \quad (14)$$

This describes the distances at which settlements are likely to be seen. Figure 6 plots dN/dr as a function of r and v. The effect of larger velocities is to scallop the density shown in Figure 3 for v=1. The reduction is not large for small velocities. In particular, for small v, say v≤.1, the distribution of visible settlements by distance is approximately given by the volume inside the light cone which has its peak at 2T/3. At the present, with T=5GY, the peak density of settlements is at 3.3GLY, so long as v<2/3. Allowing for interactions would depress the high values more than the low values. The result is a more even distribution, but

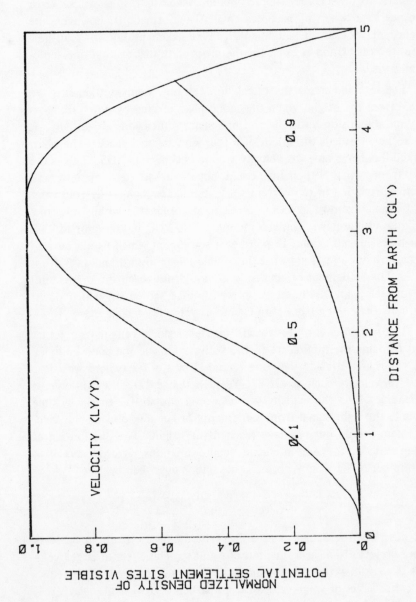

Figure 6 Effect of Expansion Velocity on Visible Settlements

with the same mode.

The total number of stars between the light cone and velocity v cone is given by

$$N = \int_0^{Tv} 4\pi kr^3(v^{-1} - 1) \, dr + \int_{Tv}^T 4\pi kr^2(T-r) \, dr \qquad (15)$$

$$= \frac{4\pi}{12} kT^4(1 - v^3).$$

The probability that no settlement is visible is thus

$$(1-p)^N = e^{-pN} = \exp(\ln p^*(v^{-3} - 1)). \qquad (16)$$

The probability that at least one settlement is visible calculated from this equation is shown in Figure 7 as a function of v and 1-p*. If we are average in emergence time, then it is very likely that at least one settlement is visible, provided v<.6. However, if v>.8, there is a significant chance (.6) that no settlement is visible.

The number of galaxies per year that may be being settled is small enough that we merely present an order of magnitude calculation. The rate of expansion of volume for a settlement originated at time t_o and whose light signal left at time t is

$$dV/dt = 4\pi v^3(t - t_o)^2. \qquad (17)$$

The rate of mass expansion for one settlement is the density of matter (kT_G) times its volume expansion rate. The total mass expansion rate is approximately the number of settlements times the average rate of mass expansion. Using $k=2 \times 10^4$ gal/GY/GLY^3, $T_G = 10$GY, and $(t-t_o)$ =2GY we find the total mass expansion to be 1 galaxy per 10 KY for p*=.99 (early) and 1 galaxy per 200 Y for p*=0.5 (middle). This is the order of magnitude for the rate at which wild galaxies are being reached in the visible volume. The process of taming a galaxy takes considerable time, though it might be as short as a thousand years (by expanding simultaneously from a thousand well-placed landing points). Thus this kind of observation is not useful for detecting settlements although it could be useful as a verification test.

To calculate the expected diameter of visible settlements, we first find the diameter of a settlement originated at (r_o, t'_o), then average over a possible t'. The visible settlement appears as an ellipsoid, because the light signal from the nearer edges left at a more advanced state of growth than that from the farther edges. The diameter at distance r_o is given by $\left|\frac{2v}{1-v^2}\right| \left|T-r_o-t'_o\right|$. This is found as the difference

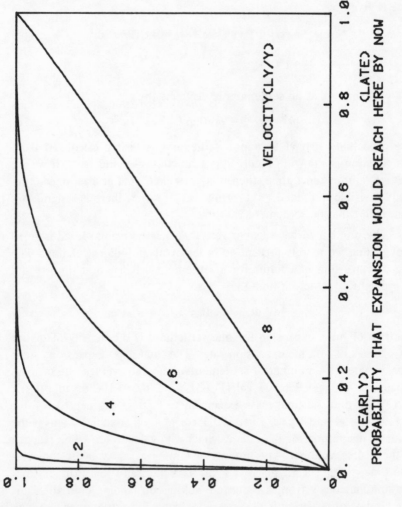

Figure 7   Observability of Settlements

in distance for the intersections of $t=T-r$ which describes the signal propagation to the origin with $t=t'_0\pm\dfrac{r-r_0}{v}$ which describes the growth of the settlement toward and away from the origin. In averaging over available t', we have again the volume between the velocity v cones and the light cones, and thus two regimes: $0<r<Tv$ and $TV<r<T$. The average t' for $0<r<Tv$ is $\frac{1}{2}\left[T-r+T-\dfrac{r}{v}\right]=T-\dfrac{r}{2}\left(1+\dfrac{1}{v}\right)$, and for $Tv<r<T$, the average is $\frac{1}{2}(T-r)$. Thus the expected diameter at distance r is

$$E \text{ diam} = \begin{cases} r(1-v) & 0<r<Tv \\ v(T-r) & Tv<r<T. \end{cases} \qquad (18)$$

Figure 8 plots Ediam as a function of r and v. This function increases to $r=Tv$ and then decreases until $r=T$. The peak value attained is $Tv(1-v)$. For $v=0.1$ or $v=0.9$, the peak value attained is about 0.5GLY.

A quantity that gives some idea of the limits to growth for a given settlement is the expected distance to the nearest other point of origination. (The nearest origination point is not necessarily visible.) The probability that the nearest origination point is in the interval $(r,r+dr)$ from the origin is the probability that no origination point is as close as r, times the probability that an origination point is in the interval $(r,r+dr)$. The expected distance to the nearest origination point is thus

$$ED = \int_0^\infty r e^{-pN(r)} p \, dN(r) \qquad (19)$$

$$=(3/4\pi pkT)^{1/3} \int_0^\infty x^{1/3} e^{-x} \, dx$$

$$=\Gamma(4/3)(3/4pkT)^{1/3}$$

$$=2.8v(-\ln p^*)^{-1/3}.$$

$$N(r) = 4/3\pi r^3 Tk. \qquad (20)$$

This distance is thus proportional to the velocity of expansion. The table at the end of the section shows the expected distance to nearest origination point for early, middle and late emergence. For middle emergence time, the expected distance to the nearest origination point would be 3GLY for $v=1$ and 0.3 GLY for $v=0.1$. The limit to growth for a newly formed settlement would be less than half this distance, because the nearest settlement would have formed sooner.

Another quantity that gives some idea of the limits to growth for a given settlement is the expected time until another settlement would

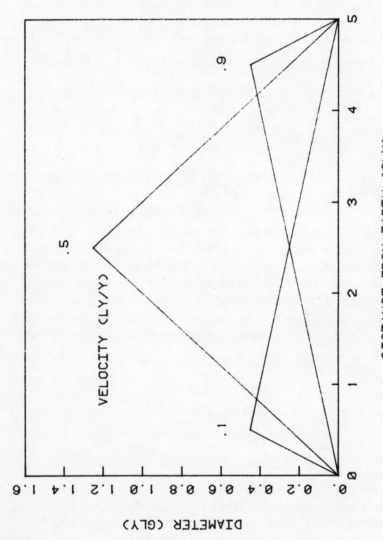

Figure 8 Expected Visible Diameter of Visible Settlements

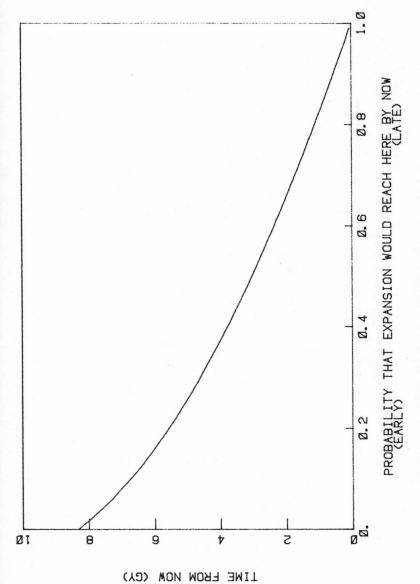

Figure 9   Expected Time until Another Settlement Reaches Here

Table 2
Summary of Model Results

	Expansion Velocity								
	v=0.1			v=0.5			v=0.9		
	.99 early	.50 middle	.01 late	.99 early	.50 middle	.01 late	.99 early	.50 middle	.01 late
prior probability that the Milky Way not colonized by now (p^*)									
probability that life originates: $p = -\ln p^* k^{-1} T^{-4} v^{-3}$ (per galaxy $= 10^{11}$ star)	$8\cdot10^{-7}$	$6\cdot10^{-5}$	$4\cdot10^{-4}$	$6\cdot10^{-9}$	$5\cdot10^{-7}$	$3\cdot10^{-6}$	$1\cdot10^{-9}$	$8\cdot10^{-8}$	$5\cdot10^{-7}$
expected time til first arrival: $ET_A = T(-\ln p^*)^{-1/2}\Gamma(5/4, -\ln p^*)$ (GY)	14	3	.05	14	3	.05	14	3	.05
expected diameter of settlements: $ED = Tv(-\ln p^*)^{-1/3}\Gamma(4/3)4^{-1/3}$ (GLY)	1.3	0.3	0.17	6.5	1.5	0.8	12	2.7	1.5
probability at least one settlement visible: $1 - \exp(\ln p^*(v^{-3}-1))$	.99996	1	1	.07	.992	1	.004	.23	.82

reach its origin (given that none has reached it yet). The probability of a settlement first reaching the origin in the interval (t,t+dt) is the product of the probabilities that none has arrived by t and that one does arrive in (t,t+dt). To account for none arriving yet, the integral is started at the current time, t'=T, rather than at t'=0. Thus we calculate

$$E(T_A) = \int_T^\infty te^{-pN(t)}pN'(t)\ dt \qquad (21)$$

$$= (12/4\pi kv^3)^{1/4}\int_{-\ln p^*}^\infty x^{1/4}e^{-x}\ dx$$

$$= T\ (-\ln p^*)^{1/4}\Gamma(5/4; -\ln p^*).$$

This velocity is independent of the expansion velocity, and is plotted in Figure 9. For early, middle, and late emergence, the formula gives 14 GY, 3.1 GY, and 50MY. Thus we can anticipate tens of millions of years of growth, even if we are late in emerging, and if we are in the middle, on the order of a thousand million years of growth. It is interesting that this approximately coincides with the fusion based mass horizon. Table 2 summarizes the results of this model.

Kardashev (1971) posed as "the Main Dilemma" that, "There is a high probability that civilization is a universal phenomenon, and yet there are no currently observed signs of cosmic activity of intelligent creatures." The explanation of the models presented here is that the other settlements are (1) very distant and (2) very hard to observe even though they completely change the nature of the galaxies in which they reside. We now review models which make different assumptions.

4. Other Models

Other major models result from various treatments of the assertions discussed in the introduction. While not all are models of growth, they present alternatives which at present are logically possible.

The *communicative* models deny that interstellar colonization is feasible (or alternatively deny that life expands to its limits). Since life is restricted to the star of its origination, any interaction is by communication. The Greenbank Formula (Sagan, 1973) expresses the number of communicative civilizations as

$$N = R_*f_pn_ef_lf_if_cL \qquad (22)$$

where R_* is the rate of star formation, f_p is the fraction of stars with planets, n_e is the number of earth-like planets per star with planets, f_l is the fraction of earth-like planets developing life, f_i is the fraction that develop intelligence, f_c is the fraction developing civilization, and L is the average life of civilizations. The assumptions behind this formula, such as constant star formation rate, are comparable to the assumptions of the multicenter growth model, save that it assumes no preemption from expanding settlements of previously formed technological life.

The communicative models allow that coherent radiation may be found from extraterrestrial sources. Detailed proposals for equipment to search for such intelligible signals have been made (Project Cyclops, 1971), and small scale searches have been made using existing radiotelescopes (Verschuur, 1973), the most recent of which searched 200 nearby stars (Horowitz, 1978).

Communicative models can either accept or deny the possibility of swarm spheres. If accepted, then there will also be individual swarm spheres visible among the stars without life. The prospect of detecting these is not favorable[*] (Sagan and Walker, 1966; Harwit, 1973).

Another model has accepted the supposition of interstellar travel, while apparently denying the possibility of swarm spheres (Kuiper and Morris, 1977). We term this model the low impact expansive model. In this model it is accepted that the galaxy is either totally uncolonized or totally colonized. The colonizers are not detectable at a distance, however, because their impact is low. This model allows, but does not require, the presence of coherent signals or direct contact from extraterrestrial civilizations.

We shall refer to any of these alternative models as low impact models, for they all suppose that intelligence makes little visible impact. The models presented in sections 2 and 3 may then be distinguished as high impact models.

5. Comparisons and discussion

We have discussed the following different models: (1) single growth center, (2) multiple growth center with expansion velocity

---

[*] It is remarkable that large and powerful settlements should be so hard to detect. Kardashev classified class I, II, and III those civilizations (we prefer the term settlement to avoid an implication of internal unit; which might control the resources of a planet, a solar system, or a galaxy. To these, the multigalactic settlements described above may be added as class IV. The first two are considered to be detectable only through narrow band coherent radio emissions. The difficulty in detecting class IV settlements, as pointed out above means that a class III settlement would be impossible to detect. This is part of the answer to S. Lem's question (23), "Why do we not see a cosmic wonder [a class III or IV civilization]?"

significantly lower than light speed and our origination not too early, (3) multiple growth center with expansion near light speed or our origination early, (4) communicative without swarm spheres, (5) communicative with swarm spheres, and (6) low impact expansive model. These models differ in their predictions of the following possibly observable phenomena: (a) swarm spheres in the galaxy, (b) coherent radiation in the galaxy, and (c) spheres of infrared galaxies in space. The following table shows the predictions.

Table 3
Model Predictions

Model	Phenomenon		
	single swarm spheres	coherent radiation	infrared galaxy spheres
(1) single center	x	x	x
(2) multiple center (low v)	x	x	p
(3) multiple center (high v)	x	x	x
(4) communicative	x	a	x
(5) swarm spheres	a	a	x
(6) low impact expansive	x	a	x

x = not allowed
a = allowed
p = probable

Note that the multiple center model with expansion near light speed is not distinguishable by the tests cited from the single center model. Nor does it seem likely to be distinguished by any observations of other life forms. Hence that model can only be rejected by the acceptance of another model.

The test for coherent radiation in the galaxy empirically distinguishes the low impact models from the high impact models. This test is asymmetric in that a positive result is easily recognized once found, but a negative result may take several hundred years to establish because of the multitude of possible frequencies and directions.

The assumptions of the high impact models are subject to constructive proof. That is, the construction of space colonies and interstellar ships can be attempted. Again a positive result is easily recognized, but a negative result requires a long period of attempts to recognize.

Each of the models presented here requires the absence of some feature in the universe. Thus the proponents of each model need to build a negative case as well as a positive case. The multicenter growth model requires that the probability of expansive life originating be below one per thousand galaxies. While an upper bound to our ignorance of this parameter may be 10^{-2} to 10^{-4} per star, no lower bound to our ignorance has been discussed. The communicative models require that one-way interstellar travel at approximately $v/c=0.1$ is impossible, or that life does not always expand. We are not aware of any published argument that this is so. There have been arguments that travel at v/c near 1 is impossible (Purcell, 1963) and Project Cyclops (1971) argued that round trip travel was impossible, but these are not sufficiently to the point. The low impact expansive model needs to argue that a swarm sphere is impossible. Assessment of the required negatives may cast a greater light on the models than attention merely to positive assertions.

The low impact and high impact models differ in their answer to the question of why the universe is wild. The low impact models give the reason that life is essentially a minor part of the universe. It might inhabit planets around every star, and yet have little impact on the evolution of the universe. The high impact models assert that the universe now appears wild only because it is in a relatively early state of development. When fully developed, intelligence is a major part of the universe, may lead to major restructuring of the distribution of matter and energy, and may impact on the long run evolution of the universe.

The communicative and expansive models differ in the prospects they offer for human growth. In the communicative models, the limits to growth lie within the next few thousand years -- either terrestrial or solar. The high impact expansive models suggest at least tens of millions of years of growth, and quite possibly a few thousand million years of growth. The low impact expansive model makes no prediction.

A possibility consistent with the high impact expansive models is that substantial probability of disaster lies ahead of us, before we could begin colonizing the stars. As we become more powerful, potential disasters may become more common. Nuclear war may be only our first such challenge. A more general problem is the possible instability of any self-defining self-creating entity. Psychologists have noted that sensory deprivation induces insanity in humans in tens of hours. When the form of intelligence is controlled by the intelligence, can the system be stable? This could represent a positive feedback effect: if the probability of technological life is below a certain level, the inability of one life form to help (if only by being competitive) another could greatly reduce the probability of expansive technological life.

The models differ in the perspective they offer on current problems such as avoiding nuclear war, finding clean energy sources, and closing materials cycles. In the communicative models these may be the final problems before settling into a steady state for some uncertain lifetime. In the high impact expansive models, they are short range problems whose solution leads to very great resource access for our descendants. Although they are short range, they may be quite serious. The high impact expansive models remind us that expansive settlements are at best rare. These short range problems may be contributors to the rarity of expansive life forms. Perhaps a greater estimate of the rewards for successful solution of these short range problems may increase the will to solve them.

REFERENCES

1. Anderson, D.M., and Banin, A., 1975, "Soil and Water and its Relationship to the Origin of Life," *Orig. Life* 6, p. 23.

2. Bahcall, N.A., 1977, "Clusters of Galaxies," *Ann. Rev. Astronomy and Astrophysics* 15, p. 505.

3. Bar-Nun, A., Bar-Nun, N., Bauer, S.H., and Sagan, C., 1970, *Science* 168, p. 470.

4. Crick, F.H.C., 1968, "The Origin of the Genetic Code," *J. Mol. Biol.* 38, p. 367.

5. Crick, F.H.C., "Extraterrestrial Life," in Sagan, 1973.

6. Crick, F.H.C., and Orgel, L.E., 1973, "Directed Panspermia," *Icarus* 19, p. 341.

7. Davis, M., "Galaxies and Cosmology," in *Frontiers of Astrophysics, 1976.*

8. Dyson, F.J., 1966, "The Search for Extraterrestrial Technology," in *Perspectives in Modern Physics*, R.E. Marshak, ed.

9. Gualtieri, D.M., 1977, "Trace Elements and the Panspermia Hypothesis," *Icarus* 30, p. 234.

10. Hart, M.H., 1975, "An Explanation for the Absence of Extraterrestrials on Earth," *Q. J. Roy. Ast. Soc.* 16, p. 128.

11. Hart, M.H., 1978, "The Evolution of the Atmosphere of the Earth," *Icarus* 33, p. 23.

12. Harwit, M., "Infrared Observations and Dyson Civilizations," in Sagan, 1973, p. 390.

13. Heppenheimer, T.A., 1977, *Colonies in Space*, Stackpole, Harrisburg, Pa.

14. Hohlfield, R.G., and Terzian, Y., 1977, "Multiple Stars and the Number of Inhabitable Planets in the Galaxy," *Icarus* 30, p. 598.

15. Horowitz, P., 1978, "A Search for Ultra-Narrowband Signals of Extraterrestrial Origin," *Science* 201, p. 735.

16. Huang, S., 1960, "Life Outside the Solar System," *Sci. Am.*

17. Jones, E.M., 1976, "Colonization of the Galaxy," *Icarus* 28, p. 421.

18. Jones, E.M., 1978, "Interstellar Colonization," *J. British Interplanetary Society* 31, p. 103.

19. Kahn, H., Brown, W., and Martel, L., 1976, *The Next 200 Years*, W. Morrow and Co., New York.

20. Kardashev, N.S., 1971, in Kaplan, S.A., *Extraterrestrial Civilizations*, NTIS.

21. Kardashev, N.S., 1973, in Sagan, 1973, p. 218.

22. Kuiper, T.B.H., and Morris, M., 1977, "Searching for Extraterrestrial Civilizations," *Science* 196, p. 616.

23. Lem, S. *Summa Technologiae*, Krakow, Wyd. Lit. , 1964.

24. Meadows, D.H., Meadows, D.L., Randers, J., and Behrens, W.W.III, 1972, *The Limits to Growth*, Universe, New York.

25. Miller, S.L., 1953, "Production of Amino Acids under Possible Primitive Earth Conditions,"*Science* 117, p. 528.

26. Miller, S.L., and Orgel, L.E., 1974, *The Origin of Life on Earth*, Prentice Hall, New Jersey.

27. O'Neill, G.K., 1977, *The High Frontier*, W. Morrow, New York.

28. Orgel, L.E., 1968, "Evolution of the Genetic Apparatus," *J. Mol. Biol.* 38, p. 381.

29. Ostroshchenko, V.A., and Vasilyeva, N.V., 1977, "The Role of Mineral Surfaces in the Origin of Life," *Orig. Life* 8, p. 25.

30. Peebles, P.J.E., and Groth, E.J., *Ap. J.* 196, p. 1.

31. Population Reference Bureau, 1976, *World Population Growth and Response.*

32. *Project Cyclops*, 1971, Report CR14445, NASA Ames Research Center, Moffet Field, Calif.

33. Purcell, E., 1963, "Radioastronomy and Communication Through Space," in *Interstellar Communication*, A.G.W. Cameron, Ed.

34. Sagan, C., 1973, *Communication with Extraterrestrial Intelligence*, MIT. Referred to elsewhere as Sagan.

35. Sagan, C. and Khare, B.N., 1971, *Science* 173, p. 417.

36. Sagan, C. and Walker, R.G., 1966, "The Infrared Detectability of Dyson Civilizations," *Astroph. J.* 144, p. 1216.

37. Sandage, A., 1972, "Distances to Galaxies, the Hubble Constant, the Friedman Time, and the Edge of the World," *Q. J. Roy. Astr. Soc.* 13, p. 282.

38. Verschuur, G.L., 1973, "A Search for Narrow Band 21cm Wavelength Signals from Ten Nearby Stars," *Icarus* 19, p. 329.

39. York, D. and Farquhar, R.M., 1972, *The Earth's Age and Geochronology*, Pergamon, New York.

5

Improving the Prospects for Life in the Universe

Michael A. G. Michaud

Abstract

The issues discussed in this symposium are not just matters of scientific research, but also of resource allocation, political decision, and societal direction. This symposium is in fact advocating a great ambition: that we use our ideas and technological powers to expand human presence and influence in the universe. The revolutionary changes in astronomy and spaceflight, and the growing acceptance of extraterrestrial intelligence, are propelling us toward such an expansion. But this is balanced by an emerging awareness of the fragility and finiteness of the earth's biosphere and of the needs and demands of the many human societies within it.

If these extraterrestrial visions are to become reality, their proponents must begin with an educational effort. The groundwork for a naturally international project needs to be laid. Large facilities in space may require not only international cooperation, but international supervision, and a further elaboration of the international legal regime for space. There is as yet no institutional framework for extraterrestrial expansion.

An attempt to expand in space may encounter limits of physical nature, motivation, self-inflicted disaster, or contact with extraterrestrial intelligence. But it may also lead to joint efforts with other civilizations to ensure the survival of intelligence in the universe.

-editor

1. Purpose of this Symposium

Many of the questions scientists explore are not just matters of scientific research, but issues of resource allocation, political decision, even fundamental societal purposes. However unintentional it may be, they frequently get on the turf of practicing social scientists like me. This topic -- the Prospects for Life in the Universe -- is a good example. It has generated some dazzling ideas. But they go far beyond scientific research.

My purpose is to suggest a humanistic overview of this vast subject. I come as a representative of the world's second oldest profession -- diplomacy. That profession can encourage a certain gentle cynicism about the likely results of human ambition. At times, it is a fifty yard line seat on human folly. An observer of international affairs may be more conscious than most of the limits of human control over events. But he also may be more aware than most of how much human activity is the result of dreams, and transcendental aspirations.

The real theme of this symposium's discussion is not the objective examination of the prospects for life in the universe, but the advocacy of a goal: to *improve* the prospects for humanity by expanding human presence and influence in the universe. That clearly is more a question of philosophy, values, and politics than it is a question of science. We are talking not about research, but about macro-engineering. We are debating whether *Homo sapiens* should become a form of expansive technological life. We are postulating a great ambition: that we use our ideas and our technological powers to shape larger and larger parts of the universe to our design.

What we are really doing in this symposium is suggesting a long-term goal for humanity. We are talking about deliberately altering the long-term expectations of human cultures to influence their future development. There may be no precedent for such an effort, save perhaps the shorter-term mobilization of populations for war.

2. Perspective on this Purpose

We need to see this proposed effort in historical, even biological perspective, and to avoid temporal chauvinism. We might begin by recalling that biological history on Earth may be understood, in part, as the spread of organisms into a broader range of environments. Perhaps the most dramatic change occurred when some forms of life migrated out of the seas on to the land. You may recall a *New Yorker* cartoon showing an amphibious creature emerging from the water on to the shore, and looking back with a smile toward a more timid colleague to say: "This is where the action is." Subsequently, some creatures developed the ability to fly in the lower atmosphere, extending the environment of life.

The distinguished American historian, William McNeill, suggested in his book *The Rise of the West* that civilized history may be understood as a series of breakthroughs toward the realization of greater and greater power. Despite the rise and fall of civilizations, there has been a certain long-term continuity in our efforts to expand our influence over our environment, to shape it to our design, to humanize the Earth. That striving toward power, toward making the world run on human time, has required increasing amounts of energy, larger scales of organization, and some degree of inhumanity. Mobilizing large-scale social efforts also has often required some sense of goal orientation, structured belief systems, even ideologies.

Having spread our species throughout this biosphere, humanized the Earth, and extended the influence of our most powerful political entities around this globe, we have encountered the limits of this planet. Having looked down on the Earth from space, we have seen our biosphere as a fragile, inter- dependent whole, and ourselves as riders on the Earth together. We in the developed world are newly sensitive to the needs and demands of other human societies, many of which have achieved political independence only since World War Two. The ever-expanding Western model of the future has been called into question.

The problems created by the paradigm of limited growth and redistribution of finite resources are very real for people in my business. We have found that our institutions are not well suited to deal with global issues, with macro-problems that extend across national boundaries, such as the energy crunch, rising prices of minerals and other basic materials, the world food problem, pollution and climatic change. We have too few generalists capable of thinking about these problems in global terms -- in terms of the species as a whole.

There is an even more fundamental consequence of accepting the idea of the limits to growth, of a world based on redistribution rather than expansion. Seeing the limits of the Earth, and of the influence of our own nation state, may already have influenced our values and our politics toward more modest expectations. We in the West may have begun a subtle transition toward a less ambitious vision of the human future. In doing so, we may be surrendering an important part of our claim to world leadership.

3. Three Revolutions

But this change in our thinking is occurring at the same time as another new current of human thought. Our time is characterized by three revolutions which are drawing the mind of Man beyond the limits of the Earth.

The new *astronomy* , made possible by bigger and better instruments, observations in non-visual wavelengths, and the lifting of instruments above the atmosphere, has shown us a different kind of universe. Gone forever are the star-studded bowl of night, rotating about the Earth or the Sun, and the stately clockwork of the Renaissance universe. Instead we find a changing, evolving cosmos, a place of violent events, of sudden births and crushing deaths. Its parts are expanding outward at reckless speeds, scattering the matter from the exploded ylem. And the structured energy of the universe is dissipating slowly into random heat and dead stars through the process we measure as entropy. This universe makes us feel not only small, but also fatalistic. It dwarfs our puny powers, making us seem irrelevant to the larger course of events, parasites on a ball of rock far from the center of anything. But the new astronomy also has shown us other worlds in unprecedented detail, stirring new dreams of alternate homes for humanity. It has revealed minerals we might consume to meet our needs, atmospheres we might change to fit our biologies. It has made our solar system a familiar place, almost within our reach.

Spaceflight is a truly new capability for the human species. For the first time in the four billion year history of life on this planet, some of Earth's creatures have the means to expand beyond this biosphere, and to seek or create others. We already have populated near-Earth space with useful machines, studied the Earth from above, and begun to inhabit space stations for months at a time. We are beginning to design larger structures for practical purposes in space, including space manufacturing facilities and satellite solar power stations. Instead of regarding space only as a new arena for exploration and technological competition, we are beginning to think about how humans might work and live there permanently, expanding the human economy and human civilization beyond the limits of the Earth. Spaceflight gives us the means to transport ourselves throughout the solar system, to build space colonies, or to found colonies on natural worlds.

The third revolution is the acceptance of the idea of *extraterrestrial intelligence*. During the past 20 years, this old idea has gained growing scientific credibility as we have found better evidence that life on Earth evolved chemically from inanimate matter, and that the raw materials of carbon-based life exist in many places in our galaxy. At the same time when we find that the universe may be indifferent to our fate, we find that it may be populated with other intelligences. This is a revolution of Copernican proportions in our conception of our place in the universe. It suggests that we might someday enter into some kind of relationship with an extraterrestrial civilization, with results that could be a boon or a disaster.

4. The Balance: Limits and Opportunity

We suffer from a dualism in our thinking. We recognize that there may be limits to our growth on Earth, and that we must share the planetary product with societies which also want to grow, At the same time, we sense a nascent consciousness of things much larger than our familiar world, new frontiers, and suggestions of an expansive future. So we hesitate about where to go from here in space. Yet our delay in exploiting this window of opportunity could close off choices for our descendants if the no-growth paradigm -- or a failure of nerve -- should come to dominate the industrial nations, and if the developing world should force a redistribution of that surplus that makes growth possible.

We are at a new stage in the growth of human power, and the expansion of Earth life into new environments. Because of our technologies, and the scales of our political and economic organizations, we now have the option of taking a conscious evolutionary step, expanding the presence and influence of humanity beyond the biosphere that evolved us -- and possibly beyond the limits that otherwise would constrain our future. We are ready to go where the action is. But we are not sure what to do with these new abilities, or what our purposes should be beyond the Earth. And we are not sure that expanding beyond the Earth is worth the cost.

We are struggling toward a change in the dominant paradigm of the human future. It is not yet clear how this struggle will come out. But we, driven by new perceptions from our astronomy and our spacecraft, have added an extraterrestrial model of the human future, with several variations. This model might be dismissed as nothing but a philosophical vehicle for technological optimism, or an outlet for human chauvinism. But it just might be an essential part of building a new cosmic context for Mankind.

Strategies for extraterrestrial growth are beginning to emerge. Arguments have been presented as to how we can use extraterrestrial sources of minerals and energy, and extraterrestrial living space, to escape the limits to growth on Earth. There is a growing, though unfocused, public interest in the extraterrestrial dimension of human existence. There are small, still poorly organized social movements behind satellite solar power, space industrialization, and space settlement. There are hopeful signs of a broadening participation in human space activities. The steps proposed may be justified by our search for the supplies that make an expanding technological civilization possible, and for new opportunities for diversity. But there is, as yet, no broad political consensus supporting these ventures. no universal attachment to the extraterrestrial paradigm. Extraterrestrial Man has not yet convinced Football-Watching Man.

5. Political Planning for a New Paradigm

To engage the human species in these macroengineering projects, to reshape and consume parts of Earth's larger environment, will require the commitment of political energy and financial resources on a scale without precedent, except in war. These commitments may require a different approach to national decisions, one that does not depend on annual budget cycles, or on a quick payback. It may be very constructive to force societies to think in larger, longer terms.

If these extraterrestrial visions are to become reality, their proponents must begin with an educational effort to spread the extraterrestrial paradigm, to mobilize public support both for the general concept of extraterrestrial expansion and for the specific steps that will start it. There must begin a gradual shift of human values to support our long-term growth beyond the Earth. It may be time to politicize space, but in a new and constructive way.

This expansion will take us into a realm beyond national jurisdiction, a vast region without boundaries that could be a natural arena for international cooperation. But few national elites share our interest in expansion, and many will suspect our motives. Large space projects such as satellite power stations or space manufacturing facilities will provoke envy, resentment, even fear; other nations may demand the right to inspect or to join in the administration of these activities, even though they do not have the financial or technological resources to participate fully. Large facilities in space may require not only international cooperation, but international supervision, and a further elaboration of the international legal regime for space.

If the Soviet Union, for example, were to build a satellite solar power station in orbit, generating gigawatts of power and capable of aiming focused beams of energy at points on the Earth or in its atmosphere, there might be in the Pentagon a certain feeling of unease. And what happens when we -- or the Japanese -- begin mining the Moon or the asteroids? According to the 1967 Outer Space Treaty, celestial bodies are beyond national jurisdiction, part of the common heritage of Mankind. A Moon Treaty has been under discussion for years; we can easily foresee a Law of the Moon conference involving interminable meetings among diplomats.

What is technologically possible may not be politically possible, unless we design for it in advance. That design may require some form of international or even supranational organization. But there is, as yet, no institutional framework for extraterrestrial Man. The United Nations Outer Space Committee provides a useful forum for discussion and consensus documents, but little more. Intelsat, or even the Alaska Pipeline consortium, may be a better model.

The internationalization of human activity in space could have a larger positive impact. Expanding our species beyond the Earth -- like the search for extraterrestrial intelligence -- defines our species by contrast with a larger environment. It makes us more conscious of our identity, and forces us to think of Humankind as a whole. The extraterrestrial adventure is potentially a shared, unifying experience for humanity.

To expand the human species beyond the Earth, to influence and control ever larger parts of the universe, is a transcendental goal for Humankind. That is something we may sorely need; a shared vision of the future can give added cohesion to civilized societies, and a grand enterprise can give a culture a sense of purpose and hope. If nations see value in the extraterrestrial effort, that outward activity might direct some of their energies away from contesting with each other on the Earth, and toward something beyond themselves. Without such an external goal, our political entities may have as their primary purpose nothing better than the periodic redistribution of wealth, status, and power, often by force. Conflict might become even more difficult to control, something we can not afford in a nuclear age. If we do not grow, if we do not join together in a shared, outward-looking human enterprise, we must face the prospect of living together indefinitely in a closed, increasingly depleted, and probably authoritarian world. Then you will need the services of my profession even more than you do now.

6. Potential Problems for an Expansionist Paradigm

If humanity does choose to grow beyond the Earth, that growth might continue for a very long time, in our solar system and beyond it. Ultimately, there are four possible limits to human growth: physical, motivation, self-inflicted disaster, and contact with extraterrestrial civilizations.

Physical limits may be the least likely to affect our future, and the easiest to overcome. One may be the speed of light, which limits the velocity of interstellar expansion. But even if that limit can not be overcome, it need not prevent interstellar migration. Even in our sparsely inhabited region of the Milky Way, stars are only about seven light years apart on the average -- a distance that can be traversed in less than a human lifetime by ships traveling at just over a tenth the speed of light. Interstellar travel would be expensive and would impose high energy requirements. But it would be accomplished by a civilization much richer than our own, with a broader base of resources, finance, and technology. A shortage of habitable planets in the solar neighborhood need not constrain us; we can build our own biospheres out of the raw materials likely to be orbiting nearby stars. If our remote descendants ever attempt intergalactic flight, the light speed limit may be a more serious problem. But if vehicles can reach high relativistic velocities, the apparent elapsed time for the travelers might not be unreasonable, due to time dilation. The ultimate physical limit to our influence on our larger environment may be the expansion of the universe, the spraying apart of the galaxies and their clusters. At some point, we might find that we did not have the energy resources or the reach in space-time to gather in more distant forms of matter and energy.

Perhaps a more immediate limit on our growth will be a loss of *motivation.* Societal values change on time scales that are relatively short compared to those of biological, geological, and cosmological evolution. We cannot predict whether or not our descendants will *want* to expand. They may choose instead to accept a no-growth paradigm, either within Earth's biosphere or within the solar system. A loss of motivation has led to the decline of human civilizations in the past. This may be a good argument for rediversifying the species, so that there will be separate cultures with different values.

A third limit to growth might be a *self-inflicted disaster* that would destroy or cripple our species. That disaster could be caused by warfare, disease unleashed by genetic experiments, ecological catastrophe, or some other cause. This is an argument for more effective social control of conflict and of dangerous technologies. It also is an argument for spreading out the species, so that not all humans would be affected by one disaster.

The fourth possible limit on our expansion is *contact with extraterrestrial civilizations.* If alien intelligences have evolved elsewhere in the universe, some may have developed technical civilizations long before we did. Some of those in turn may have chosen to expand, occupying growing regions of their galaxies. Many star systems may have been occupied, and these expansionist species may have begun manipulating their macroenvironments. If such civilizations exist in this galaxy or in galaxies nearby, they imply a limit on our expansion, at least in one direction. Even if our expansion were to bring us into contact with less powerful civilizations, we might choose to limit our expansion in that area for ethical reasons.

7. Implications of the Expansionist Paradigm

Other intelligent species also may have passed through our present stage. Confronted with the yawning emptiness of the space about them, the apparent indifference of a universe expanding away from them and losing its structured energy, some civilizations may suffer a loss of nerve, and turn inward on themselves, soothing the pain with metaphysical balm. But others may be motivated by a desire to increase their influence in the universe, to broaden their options for survival. They too may begin to expand away from the biospheres that evolved them, toward some transcendental goal. Their expansions may not be uniformly successful; they may be slow, halting, unevenly distributed, and may fail at any stage for philosophical and political reasons.

At some time in galactic history, some of these expanding spheres of intelligence -- these noospheres -- may touch through communication or direct contact. The results could range from disastrous conflict to cultural and intellectual symbiosis, even synthesis. It may be that each expanding species regards survival in the universe as a zero-sum game. Freeman Dyson may be right in saying that we are most likely to encounter first a species in which technology is out of control, a technological cancer spreading through the galaxy. If so, our noisy, disorganized, and technologically primitive species may have a limited future after all.

At least some civilized beings may conclude that they should explore the possibilities for cooperation before embarking on unilateral actions that could limit the future of others. If one species can not impose its purposes on another, it may need to persuade instead. Galactic civilizations may even need the functional equivalent of diplomats.

Instead of contesting for entropy control, the intelligences of the universe -- including this local example -- might see greater rationality in sharing their perceptions and their ideas, their values and their purposes. They may see a need for a transcendental goal shared among intelligent species. They may find that all intelligent species share one goal: the survival of intelligence itself, and the extension of its influence in and over the universe. They might reach out to each other to create a higher level of organization among the most aware and highly advanced products of cosmic evolution.

8. Conclusion

So we return to the basic question: can intelligence have an impact on the universe, or is it a minor phenomenon, an unimportant by-product of cosmic evolution? One suspects that the same question has been asked in symposia on distant worlds, in languages and modes of thought we could not understand. Today's discussion suggests that intelligence can have an impact, but only if each intelligent species is prepared to make massive political decisions, to alter its values and its expectations to support an extrabiospheric model of its future, and to commit itself to societal purposes that may reach far beyond the lives of individuals.

Among humans, our generation is the first to have this choice. It may be up to us to prove that intelligence armed with technology has long-term survival value. It may be up to us to take the first outward steps in an expansion that will consume part of our species' energy for thousands, even millions of years. And it may be up to us to prove that expanded consciousness is not a tragedy, but a joy.

If our own expansion is successful, it may lead to the spread among our descendants of the vice the ancient Greeks called *hubris*. These star people may conclude that they are potentially masters of the Galaxy, God-like creatures with a special mission in the universe. That new anthropocentrism must be balanced with an extraterrestrial ethos that includes a profound respect for other living things, for other intelligences, and even for the most beautiful works of inanimate evolution. There may be no room in the universe for human chauvinist pigs.

As we search for alien intelligences, we must prepare ourselves intellectually and culturally for possible contact, so that it will neither irrevocably damage us or unnecessarily provoke them. Contact would require us to think as a species; responding implies cultural and political changes that would allow us to speak as one, perhaps through a new global institution.

If we do not find other minds -- if we truly are alone -- it will be our task to assure the survival of intelligence in the universe. That may be our ultimate moral task. It may be up to us to answer the question posed earlier in this symposium: is the universe pointless? We may have to answer by supplying the point of existence ourselves. That is task enough.

Index